TAHOE
WILDFLOWERS

A MONTH-BY-MONTH GUIDE TO
Wildflowers in the Tahoe Basin
and Surrounding Areas

BY LAIRD R. BLACKWELL

FALCONGUIDE®

GUILFORD, CONNECTICUT
HELENA, MONTANA
AN IMPRINT OF THE GLOBE PEQUOT PRESS

*A*FALCONGUIDE®

Text design by Sue Murray
Maps by Multi-Mapping Ltd. © Morris Book Publishing, LLC
All photographs © Laird R. Blackwell unless otherwise noted.

Library of Congress Cataloging-in-Publication Data
Blackwell, Laird R. (Laird Richard), 1945-
   Tahoe wildflowers; a month-by-month guide to wildflowers in the Tahoe Basin and surrounding areas / Laird R. Blackwell.—1st ed.
      p. cm.—(A Falcon guide)
   Includes bibliographical references and index.
   ISBN-13: 978-0-7627-4369-8
   ISBN-10: 0-7627-4369-7
   1. Wild flowers—Tahoe, Lake, Watershed (Calif. and Nev.)—Identification. 2. Wild flowers—Tahoe, Lake, Watershed (Calif. and Nev.)—Pictorial works. 3. Tahoe, Lake, Watershed (Calif. and Nev.) I. Title.
   QK142.75.B53 2007
   582.1309794'38—dc22

2006022285

Manufactured in China
First Edition/First Printing

CONTENTS

MAP

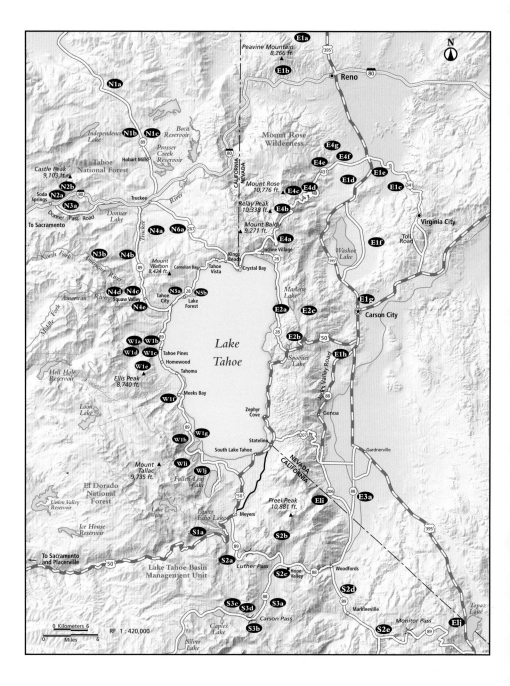

INTRODUCTION

One of the wonderful benefits of being in the mountains where elevation shifts rapidly in short distances is the opportunity to change seasons rapidly as well with just a hike or short drive. To wildflower lovers this is an especially appreciated gift, for in many months of the year, with a bit of a climb or descent, we can move back or forward several months in the flower season. This "seasonal mobility" is one of the great attributes of Tahoe for wildflower enthusiasts, for we can experience more than 6,000' of elevation change within 25 miles of the lake. Lake Tahoe (at about 6,200') is ringed with mountains that consistently crest at 2,000–3,000' above the lake and, in a few cases, reach over 4,500' above those deep blue waters. To the north, in the corridor along the Lower Truckee River as it flows north out of the lake, there are miles of mostly flat landscape gradually descending to about 5,900', while to the east on the eastern escarpment of the Carson Range that rises thousands of feet above the lake's eastern shore, the land descends abruptly to the high Great Basin desert in Washoe Valley and Carson Valley at 4,600–5,000'. To the west and south it's all uphill from the lake—reaching the Sierra crest to the west at 8,000–9,000' and peaking to the south at 10,881' atop Freel Peak, Tahoe's highest summit.

This means joy for flower lovers, for in March (in some years even in February) when snow is heavy in the Tahoe Basin, we head east to the sandy flats of Carson City to find the tiny, graceful, white flowers of the aptly named spring whitlowgrass and a few other early spring bloomers, and we smile, knowing that the flowers are returning once again. And every year we wait eagerly for May and the great lakes of lush, blue camas lilies that flood the low-lying meadows north of Lake Tahoe, bringing the vibrant joys of spring in their wake. Later, in July, we stand surrounded by incredible, shoulder-high, rainbow gardens near such places as Winnemucca Lake to the south, below Tamarack Lake to the east, in Paige Meadows to the west, and on Castle Peak to the north, and our hearts race in midsummer exhilaration with the bursting blooms. In those brilliant summer days of July and August, we can climb with the flowers to Tahoe's highest point, the summit of Freel Peak, and exhilarate in the "sky gardens" adorning the scree and talus here "on top of the world" as we look out over the entire Tahoe Basin stretched beneath us.

And every year, we eagerly await the late bloomers of September and October even though we know that they will be the last hurrah of the blooming season, for they offer a spectacular farewell to the flower year—in the volcanic cliffs above Pole Creek, gentian

and fuchsia usher the year out with "loud" blues and scarlets, while along gravelly road edges in such places as Spooner Summit, Carson Pass, and Monitor Pass, rabbitbrush, goldenrod, and Wright's buckwheat sound softer tones.

For those of us who live here or who visit frequently, each of these months has its special flavor and character, created by the flowers that adorn it. And what a glorious gift these flower months are in this mountain Shangri La, for we are blessed with an incredibly rich and varied "bouquet": The rich floras of the Great Basin to the east, the Sierra foothills to the west, the volcanic Cascades to the north, and granitic Yosemite to the south all send out their threads to weave a magnificent floral tapestry here in Tahoe.

How to Use This Book

You can use this book to know where to go in the Tahoe area any month from February or March through September or October to see special wildflower treasures—spectacular, ˙ ʌˍ ʌˌ ˍ.d heautiful individuals. The common, the not-so-common,
for you.

es from Kyburz Flat (25 miles north of Lake Tahoe) to
5 miles south of Lake Tahoe) and from the Sierra
to Topaz Lake and Peavine Mountain (along the
eral miles southeast and northeast respectively of the
covered ranges from the Great Basin desert at about
' summit of Freel Peak—well over a vertical mile. The
n all its range and diversity, is within an hour's drive
ie lake.

oduction, you will find a map showing the location of
oe where you can find glorious wildflower displays
utiful, common, or unusual flowers. The key to the map
ription of each place, lists some of its special flowers,
ctions.

gh September or October (or, in some years, even Novem-
lere in the Tahoe Basin or in the Great Basin valleys and
slopes to the east. ʌ ˍ. ʌ . most flowers will be in bloom for several weeks, and since most species occur across a fairly broad elevational range, you will probably be able to find most species in bloom somewhere in the Tahoe area across at least 2 or 3 months.

Deadman's Creek

Topaz Lake

The main text of the book is organized by month, from February through September. For each month some of the most interesting or spectacular flowers that usually start blooming during that month are briefly described and illustrated. (Blooming times can vary by 2–3 weeks or in some years even more depending on the winter and spring weather, and not all flowers will bloom every year.) Within each month the flowers are sequenced by color and, within color, by family (scientific name) and species alphabetically. For each flower several places where you can find it—either a few plants in the case of infrequent or rare flowers or especially glorious displays in the case of more common ones—are indicated. Although each flower (description and photo) is presented only in the month in which you are likely to first find it, most flowers (especially those with a wide elevational range) can be found blooming in several months, so in most cases the descriptions will indicate several places and months to look for it. If you are looking for a particular species but don't find it described under the month you are looking at, you might want to check other months or check the indexes, which include the 200 species described in this book as well as more than 300 additional species listed. For the described species, the page num-

Red Lake Peak at the south end of Lake Tahoe

ber is indicated; for the additional species, one of the best places and times to find it is indicated.

The descriptions of the 200 featured flowers include the plant's typical habitat, its geographical and elevational distribution in the Tahoe area ("low" is below 6,500', "mid" is 6,500–9,000', and "high" is above 9,000'), distinguishing characteristics, and especially good places (with months and approximate elevations) to find it in bloom.

Key to the Places with Driving and Walking Directions
(All hiking and walking distances are ONE WAY)

The map on p. iv shows the location of each of the 56 places featured in this book where wildflowers may be found. This key identifies and briefly describes each place and lists some of the special flowers and flower gardens you are likely to find there. **Special gardens** indicates that, in my opinion, the displays of this flower along this walk or trail are among the most spectacular and profuse of any place in the Tahoe area. An **asterisk** indicates that this flower can be found in very few other places in the Tahoe area. **Flowers of note** indicates flowers of special interest (those that are found in few, if any, other places in the Tahoe area) that are found only occasionally on this walk or hike, not in great masses or gardens. For the months to find these flowers or gardens in full bloom, check the flower description.

The place descriptions are organized by location in relationship to Lake Tahoe: North (from Castle Peak northwest of the Lake to Kyburz Flat to the north); West (along the west shore of the lake and a few miles westward up to the crest of the Sierra Nevada); South (from Carson Pass south of the lake to Monitor Pass to the southeast); and East (along the east shore of the lake and several miles eastward up and over the crest of the Carson Range to Washoe Valley and Carson Valley in the Great Basin desert from Peavine Mountain in the north to Topaz Lake in the south).

East of Lake Tahoe

E1: Route 395 from Peavine Mountain (north of Reno) south to the intersection with Route 89 just south of Topaz Lake

E1a: Peavine Mountain from the northeast—strenuous, 4-mile hike to summit (5,500–8,266') March–July

Driving: From the intersection with Route 80 in Reno, drive about 9.5 miles north on Route 395, turn left on Red Rock Road, turn right on Virginia Street. After about 1.4 miles turn left on the dirt road at the crest of the hill, cross the RR tracks, drive up the road until it starts to get steep.

Walking: Park and walk up the very steep road, staying on Route 642 when you have choices. Walk as far as you like toward the summit.

special gardens: grand collomia, porterella, sego lily, aspen onion, bitterroot, Anderson's milkvetch, California hesperochiron, Anderson's larkspur, daggerpod, big-headed clover, Applegate's paintbrush, Copeland's owl's-clover, silver lupine, showy penstemon, low cryptantha*, tobacco brush, ochre-flowered buckwheat

flowers of note: narrow-leaf phacelia, apricot mallow, northern bog violet, Kelley's tiger lily

E1b: Peavine Mountain from the southeast—strenuous, 5-mile hike to summit (5,000–8,266') March–July

Driving: From the intersection with Route 80 in Reno, drive about 2 miles west on Route 80, turn right on Keystone. After about 1.6 miles turn left on McKarren, take the first right on Victory, turn immediately left then right into the East Keystone trailhead.

Walking: Walk up the road through the canyon, take Route 655 at the buildings, curve left up the steep hill. From here you can see the summit (and towers) way above you—there are many ways to get there, which will be apparent.

special gardens: white layia, soft lupine*, altered andesite buckwheat*, purple sage, bitterroot, golden phacelia*, fiddleneck, common purple mustard, Copeland's owl's-clover, California poppy, checkermallow, desert paintbrush, shaggy milkvetch, blepharipappus

flowers of note: purple nama, golden currant, apricot mallow

E1c: Old Geiger Grade (Toll Road) and Geiger Grade (Route 341)—strolls along road (4,900–6,200') March–July

Driving: **For Old Geiger Grade,** from the intersection with rte, 431 (Mt. Rose Highway), drive about ¾ mile east on Route 341, then right on Toll Road at the stoplight. After about 2 miles the road turns to rough dirt. Continue on the main road (left fork) as far as you like, but the road can be impassible at times, so you might want to park and walk. **For Geiger Grade** (Route 341), continue east on Route 341 past the Toll Road turn. You could drive all the way to Virginia City.

Walking: For Old Geiger Grade and Geiger Grade, explore along the road edges. If you decide to park at the bottom of Old Geiger Grade and walk the road, it's a long walk if you go all the way to the top and back.

special gardens: altered andesite buckwheat*, common purple mustard, cushion desert buckwheat, Beckwith violet, dwarf onion, Sonne's desert parsley, Hooker's balsamroot, bitterroot, Anderson's clover, prickly poppy, showy penstemon, Pursh's milkvetch, cobweb thistle, blazing star

flowers of note: Kellogg's onion, yellow club-fruited evening-primrose

---

E1d: Ophir Creek trail—strenuous, 6-mile hike to top (5,200–8,700') April–August

Driving: **From the north,** from the intersection with Route 431 (Mt. Rose Highway), drive about 8.6 miles south on Route 395, turn right on Bower's Mansion Road. After about .4 mile turn right on Davis Creek Park road and park near the Ophir Creek trailhead. **From the south,** from the intersection with Route 50 West just south of Carson City, drive about 13.6 miles north on Route 395 (Carson Street to its end where it enters the freeway), turn left on Bower's Mansion Road, and proceed as described above.

Walking: In spring you might want to hike the 1st mile or so of the trail; in summer, you could start at the north end of this trail (in Tahoe Meadows) and hike down. You'd need to have a car waiting at the Davis Creek Park end. It's a long hike with some trail-finding challenges, so be prepared.

special gardens: yellow-or-purple monkeyflower (red), rayless daisy, greenleaf manzanita, white layia, bitterbrush

flowers of note: tiny evening-primrose

E1e: Steamboat Springs and Wahoe Valley—strolls along road (4,600–5,000') March–July

Driving: **From the north,** from the intersection with Route 431 (Mt. Rose Highway), drive about 2 miles south on Route 395, turn left on Rhodes Road and almost immediately left on Steamboat Springs Road, and drive past the buildings. When the road takes you back to Route 395, turn left and continue south on Route 395 for about 8 miles to where the valley opens up, continue another 5 miles or so through the valley to the west of Washoe Lake. **From the south,** from the intersection with Route 50 West just south of Carson City, drive about 8 miles north on Route 395 (Carson Street to its end where it enters the freeway) to where the valley opens up. Then continue about another 13 miles, turn right on Rhodes Road, and continue as described above.

Walking: Explore the road edges, being respectful of private property and fences.

special gardens: tansy-leaf evening-primrose, wild iris, crisp thelypodium*, two-horned downingia*, yellow club-fruited evening-primrose*, Steamboat Springs buckwheat*, yellow-or-purple monkeyflower (red), rayless daisy, common yellow monkeyflower, desert peach, blue sailors, wallflower, Torrey's lupine

flowers of note: sacred datura, moth mullein

E1f: Deadman's Creek—easy, ½-mile walk (5,000–5,200') March–September

Driving: **From the north,** from the intersection with Route 431 (Mt. Rose Highway), drive about 15 miles south on Route 395, take exit 42, and turn left on East Lake Boulevard. After about 3 miles park in the turnout on the right at Deadman's Creek trailhead. **From the south,** from the intersection with Route 50 West just south of Carson City, drive about 7 miles north on Route 395 (Carson Street to its end where it enters the freeway), take exit 42, and continue as described above.

Walking: Follow the trail as it heads east then curves back west up to the gazebo. Explore the rocky west-facing slope below the pergola.

Tamarack Lake

special gardens: wild rose, common yellow monkeyflower, desert peach, Veatch's stickleaf, Anderson's milkvetch, sainfoin*, spreading phlox, daggerpod, arrowleaf balsamroot, rabbitbrush, cobweb thistle, aspen onion

flowers of note: desert gooseberry, Washoe phacelia

E1g: Carson City—strolls in field (4,700') February–May

Driving: **From the north,** from the intersection with Route 431 (Mt. Rose Highway), drive about 18 miles south on Route 395 into Carson City, take the College Parkway exit to head east on College, then take the first left on Research Road, turn right on Old Hot Springs Road, and turn left on Wedco Way. Park near the end of the road. **From the south,** from the intersection with Route 50 West just south of Carson City, drive about 5 miles north on Route 395 (stay on Carson Street), turn right on College Parkway. After about 1 mile turn left on Research Road and continue as described above.

Walking: Explore the field to the west. This field is on airport property, so explore carefully and leave no litter.

special gardens: Nevada lomatium, bur buttercup, dwarf onion, spring whitlowgrass, white layia, desert peach, Hooker's balsamroot, tiny evening-primrose*, yellow-or-purple monkeyflower (yellow)*, slender pop-corn flower, Anderson's milkvetch, silver lupine, shaggy milkvetch

E1h: Jack's Valley Habitat Management Area—strolls along dirt road (4,900–5,200') April–June

Driving: **From the north,** from the intersection with Route 50 West just south of Carson City, drive about 1½ miles south on Route 395, turn right on Jack's Valley Road. After about 1¾ miles turn right onto the dirt road into Jack's Valley Habitat Management Area (sign). **From the south,** from the intersection with

Route 88 in Gardnerville, drive about 9½ miles north on Route 395, turn left on Jack's Valley Road and continue as described above.

Walking: Drive or walk up this dirt road as far as you like and explore the sagebrush slopes.

special gardens: bitterbrush, arrowleaf balsamroot, desert paintbrush

---

E1i: Faye-Luther trail—strenuous, 2-mile hike partway up trail (4,850–6,000') May–August

Driving: **From the north,** from the intersection with Route 50 West just south of Carson City, drive about 1¼ mile south on Route 395, turn right on Jack's Valley Road, continue about 18 miles south through Genoa, turn right into the Faye-Luther trailhead parking area. **From the south,** from the intersection with Route 88 in Gardnerville, drive about 6½ miles west on Route 88, turn right on Route 206 (Fairview Lane). After about 1¾ miles turn left into the Faye-Luther trailhead parking area.

Walking: Hike up the trail behind the sign, bear left at the first fork before you reach the woods. Stay on the main trail heading south (there will be some forks heading west into the woods), which will eventually curve west and climb up toward the Carson Range. You could continue several miles steeply up toward Freel Peak, but it's quite a climb and there are much easier ways to get to that wonderful mountain, so I suggest just hiking the first couple miles on this trail.

special gardens: sego lily, tobacco brush, yellow-or-purple monkeyflower (red), desert peach, snowplant, dwarf phacelia

flowers of note: purple milkweed, narrow-leaf phacelia

E1j: Topaz Lake—strolls up wash (5,100') March–July

Driving: **From the north,** from the intersection with Route 88, drive about 25 miles south on Route 395, park in one of the dirt turnouts (wide shoulders) on the right just beyond the Topaz Lake sign. **From the south,** from the intersection with Route 89 just south of Topaz Lake, drive about 2 miles north on Route 395, and park on the left as described above.

Walking: Walk up the draw and onto the ridge.

special gardens: long-leaf phlox, Applegate's paintbrush, rayless daisy, white-flowered keckiella*, prickly poppy, apricot mallow*, blazing star, spring whitlowgrass

flowers of note: tufted evening-primrose, narrowleaf milkweed

---

E2: Route 28 from Incline Village south to intersection with Route 50, then east toward Route 395. From the north, driving directions are from the intersection with Route 431 (Mt. Rose Highway). **From the south,** they are from the intersection with Route 50 near Spooner Summit.

E2a: Prey Meadows—easy, 1½-mile walk down to meadow (6,900–6,700') May–September

Driving: **From the north,** drive about 10½ miles south on Route 28, park in the turnout on the right just before the green-gated Forest Service, paved road. **From the south,** drive about 2½ miles north on Route 28 and park on the left as described above.

Walking: Walk north down the paved road, which will turn to dirt after about ¼ mile. At the first fork take the path down to the left and switchback down to the meadow.

special gardens: alpine shooting star, snowplant

flowers of note: twayblade, baneberry

E2b: Spooner Lake—easy 2-mile loop walk (7,150–7,000') May–October

Driving: **From the north,** drive about 13 miles south on Route 28, turn left on Route 50, after about ¾ mile park in the paved lot on the left at the Spooner Summit trailhead. **From the south,** drive about ¾ mile east on Route 50 and park as described above.

Walking: Walk north on the Tahoe Rim trail. After less than 50 yards, take the left fork signed SPOONER LAKE ACCESS TRAIL down to the lake. Circle the lake and return up the trail you descended.

special gardens: peony, Jessica's stickseed, snowberry, rabbitbrush, pennyroyal, duck potato

E2c: Snow Valley Peak/Marlette Lake—11-mile loop hike (7,150–9,000') June–October

Driving: Driving and parking directions are the same as for E2b above.

Walking: At the first fork on the Tahoe Rim trail, stay right on the main trail. After about 4 miles continue straight at the sign SNOW VALLEY PEAK—1½ MILES. Just before the peak, turn left down the dirt road and intersect the dirt road to Marlette Lake. Turn right and walk to the lake. On your return stay on the dirt road you're on all the way back to Spooner Lake, circle the lake, and climb back to your car.

special gardens: monkshood, Anderson's thistle, Wright's buckwheat, great red paintbrush

E3: Route 88 from Gardnerville west to Woodfords

E3a: Gardnerville to Woodfords—strolls along road (4,700–5,500') May–July

Driving: From the intersection with Route 395, drive about 13¾ miles west to Woodfords.

Walking: Wander in the gravelly and grassy flats along the road, especially near Woodfords.

special gardens: California poppy, blue sailors, cobweb thistle

flowers of note: bouncing bet

E4: Route 431 (Mt. Rose Highway) from Incline Village east to Reno. From the west, driving directions are from the intersection with Route 28 in Incline Village. From the east, driving directions are from the intersection with Route 395 south of Reno.

E4a: west-side bowl toward Diamond Peak—easy ½-mile walk (7,800') June–August

Driving: **From the west,** drive about 3¾ miles east on Route 431 and park in the dirt turnout on the right near the two vertical stakes. **From the east,** drive about 26 miles west on Route 431 and park on the left as described above.

Walking: Follow the faint trail into the woods. Soon you'll emerge at the edge of a south-facing bowl, follow the trail left as it contours east around this bowl. After about ½ mile you will encounter a large aspen grove, at which point you might want to head down to the meadow.

special gardens: broad-leaf lupine, great red paintbrush, Jessica's stickseed, scarlet gilia

E4b: Tahoe Meadows—strolls (8,600–8,700') June–August

Driving: **From the west,** drive about 6½ miles east on Route 431 to the large, open meadows. Park on either side of the road on the wide shoulders. **From the east,** drive about 23 miles west on Route 431 and park as described above.

Walking: Stroll out into the meadows on both sides of the road.

special gardens: meadow penstemon, marsh marigold, alpine shooting star, white crest lupine, bull elephant's head, little elephant's head, bog saxifrage, Bigelow's sneezeweed, Sierra rein orchid, Brewer's lupine, corn lily, primrose monkeyflower, water plantain buttercup

flowers of note: white ladies-tresses

E4c: Mt. Rose summit—strenuous, 5-mile hike (8,900–10,776') June–September

Driving: **From the west,** drive about 8 miles east on Route 395 and park in the paved lot on the left signed MT. ROSE TRAILHEAD. **From the east,** drive about 21½ miles west on Route 395 and park on the right as described above.

Walking: Head up the trail to the west behind the kiosk. At the first fork in less than 20 yards, head right where it says HIKERS ONLY. After about 2½ miles, you will pass a waterfall on your left and intersect the trail to the summit. Head right on this trail up a narrow canyon, then bear right at the sign MT. ROSE SUMMIT—1 MILE. The last ¼ mile of this trail is across rocky talus slopes above timberline.

special gardens: Lewis monkeyflower, glaucous larkspur, corn lily, alpine penstemon*, alpine gold*, blue flax, showy penstemon, rock fringe, marsh marigold, alpine paintbrush, Brewer's lupine, spreading phlox, white crest lupine, yellow or purple monkeyflower (red)

flowers of note: alpine buttercup, woody-fruited evening-primrose, showy polemonium, alpine gentian

E4d: Tamarack Lake—easy but steep ¼-mile walk (8,700–8,800') June–September

Driving: **From the west,** drive about 9 miles east on Route 431. Shortly after you begin to descend toward Reno, park in the large paved turnout on the right. **From the east,** drive about 21 miles west on Route 431 and park in the turnout on the left as described above.

Walking: Cross the highway and walk up the hill past the sign NO MOTORIZED VEHICLES and follow the faint-use trail to the lake.

special gardens: glaucous larkspur, Brewer's lupine, wandering daisy, crimson columbine, spreading phlox, dwarf phacelia, corn lily

E4e: Galena Creek—4-mile hike and scramble (7,500–9,000') June–October

Driving: **From the west,** drive about 13 miles east on Route 431, turn left into the dirt turnout opposite the Highway Maintenance Station at the switchback. **From the east,** drive about 17 miles west on Route 431 and park on the right as described above.

Walking: Walk west along the faint trail, which soon drops down to the creek and a log crossing. Walk/scramble as far as you like on either side of the creek. If you make 4 miles, you will intersect the Mt. Rose Summit trail.

special gardens: Lewis monkeyflower, primrose monkeyflower, dwarf phacelia, Bigelow's sneezeweed, squaw carpet, greenleaf manzanita, pussypaws

flowers of note: woody-fruited evening-primrose, Indian blanket

E4f: Thomas Creek lower trailhead—easy, 1-mile walk (6,000–6,500') March–June

Driving: **From the west,** drive about 18½ miles east on Route 431, turn left on Timberline Drive. After about 1 mile, cross the bridge over Galena Creek and continue straight on the middle of 3 forks for about ¹⁄₁₀ mile and park in the paved lot for the Thomas Creek trailhead. **From the east,** drive about 11 miles west on Route 431, turn right on Timberline Drive, and continue as described above.

Walking: Walk along the upper dirt road to the north; when it bends to the left, walk straight into the large, open meadow or continue along the road as it heads up the slope.

special gardens: cushion desert buckwheat, Beckwith violet, Sonne's desert parsley, star lavender, long-leaf phlox, slender woodland star, Pursh's milkvetch, single-stem senecio, dwarf onion, naked broomrape, big-headed clover, sand corm, blepharipappus, Nevada lomatium

flowers of note: Lassen clarkia, desert buttercup, yellow violet

E4g: Thomas Creek upper trail—strenuous, 5-mile hike (6,800–9,000') June–October

Driving: Driving directions are the same as for E4f above, except that at the fork just across the bridge, take the left fork and drive about 2½ miles west on this rough, dirt road (includes a shallow creek crossing, so you might not want to try this after rains). Park just before the closed Forest Service gate.

Walking: Walk up the dirt road, which after a mile or so will reach the Thomas Creek trailhead sign. Continue up the trail, which will eventually cross the creek and climb up to a ridge, then contour around a large, open bowl. When you reach a fork in the trail, take the left fork, which shortly reaches a large, wet meadow.

special gardens: wild iris, aspen onion, horse-mint, sego lily, checkermallow, Anderson's thistle, bog wintergreen, Bigelow's sneezeweed, Applegate's paintbrush, snowberry, blue-eyed grass

flowers of note: dwarf chamaesaracha, orange mountain dandelion

North of Lake Tahoe

N1: Route 89 north of Truckee

N1a: Kyburz Flat—stroll (5,900') April–July

Driving: From the intersection of Route 267 (which turns into Route 89 north of Route 80) and Route 80 just east of Truckee, drive about 12 miles north on Route 89 North, turn right on dirt road signed KYBURZ FLAT. After about 1 mile, park in the dirt turnout on the right just before the bridge.

Walking: Wander out into the large, open meadow and/or circle the pond.

special gardens: porterella, yellow pond lily, California hesperochiron, northern suncup, dwarf phacelia, single-stem senecio, tansy-leaf evening-primrose, pussypaws

N1b: Sagehen Creek west—stroll (6,200') April–July

Driving: From the intersection of Route 267 (which turns into Route 89 north of Route 80) and Route 80, drive about 8 miles north on Route 89 North, park in the dirt turnout on the right just after the bridge crossing Sagehen Creek.

Walking: Walk southwest across the road and back across the bridge, then head to your right, follow the faint use trail through a short stretch of woods and then out into the wet meadow. Use the logs to avoid disturbing the wet ground.

special gardens: marsh marigold, bog saxifrage, bull elephant's head, crimson columbine, star lavender, dwarf phacelia, common yellow monkeyflower, blue-eyed grass

flowers of note: northern bog violet

N1c: Sagehen Creek east—easy, 2-mile walk (6,200–5,900') April–August

Driving: Driving and parking directions are the same as for N1b.

Walking: Do not cross the road, but instead walk along the trail heading east from the parking area into the woods and along the north side of Sagehen Creek. You will eventually reach a huge, open meadow, and, at its eastern edge, the California Native Plant Society fenced wildflower enclosure just west of Stampede Reservoir.

special gardens: camas lily, water plantain buttercup, three-bracted onion*, alpine shooting star, Sierra rein orchid, bog saxifrage, squaw carpet, bistort, wild rose

flowers of note: white ladies-tresses, Bacagalupi's downingia

N2: Route 80 west of Truckee

N2a: Castle Peak west—strenuous, 4½-mile hike and scramble (7,200–9,115') June–August

Driving: From the intersection with Route 89 south in western Truckee, drive about 8½ miles west on Route 80, exit at Boreal Ridge Road (1 exit west of the Rest Area), turn right on the paved spur road, park along the road where it turns to dirt.

Walking: Hike north up the dirt road, into meadows, and up the Pacific Crest Trail. You might want to scramble (carefully!) up toward the ridge to your east. (You can drive farther up this dirt road to shorten your hike if you prefer.)

special gardens: Drummond's anemone, Sierra primrose, pretty face, creek dogwood

N2b: Castle Peak east—strenuous, 4-mile hike (7,200–9,103') June–August

Driving: Driving and parking directions are the same as for N2a above.

Walking: Walk up the faint trail heading southeast into the woods from the dirt parking area on the right side of the spur road. You will walk behind the pond and the rest area and join the main trail (initially paved) heading east toward Castle Peak and Summit Lake. After a mile or so, turn left at the junction toward Warren Lake. After a couple of miles with some steep climbing and a couple seep areas, you will reach a ridge off of which the trail descends north toward Warren Lake. Instead of descending, cross-country along the ridge to your left (west). You will eventually reach some springs and lush gardens on the south-facing slope, then the rocky summit area. (When you return and pass behind the rest area, be careful not to miss the faint trail branching off to your left back to the car—it is easy to miss this turn and continue on the main trail. If you start bearing right [north], you have missed it.)

special gardens: alpine shooting star, corn lily, bull elephant's head, Copeland's owl's-clover, Drummond's anemone, Sierra primrose, tiger lily, pink gilia, primrose monkeyflower, grand collomia

N3: Route 40 (Donner Pass Road) west of Truckee

N3a: Donner Lake to Soda Springs—strolls along road (6,000–7,200') May–September

Driving: From the intersection of Route 89 south and Route 80 (freeway ramp) in western Truckee, drive about 4¼ miles west on Route 80, take the Donner Lake exit. After about 1½ miles you will reach Donner Lake. Turn right on Donner Pass Road and climb up to Soda Springs.

Walking: Explore the road edges, especially the gravelly flat on the right at about 2 miles west of the west end of Donner Lake where the road begins to climb.

special gardens: Sierra stonecrop, steershead*, azure penstemon, Leichtlin's mariposa lily, pretty face, dwarf Sierra onion, fuchsia, mountain pride

flowers of note: Washington lily, northern bog violet

N3b: Ridge Route from Donner Summit to Squaw Valley—15-mile one-way hike (7,100–8,600–6,200') June–August

Driving: Same driving directions as for N3a above except continue up the Donner Pass Road until you reach the summit, turn left at the Alpine Skills Institute, take the first left to the Pacific Crest Trail sign MT. JUDAH—1½ MILES.

Walking: Walk the Pacific Crest Trail south all the way to Squaw Valley, take the left fork down Shirley Canyon on the north side of Shirley Creek. Just after you pass behind the fire station, take the short spur trail to your right along the east side of the Fire Station out into the parking lot. You will need to have a car waiting for you here.

special gardens: Sierra primrose, mule ears, pennyroyal, Lobb's buckwheat

N4: Route 89 from Truckee south to Tahoe City. From the north, directions are from the intersection with Route 80 in western Truckee. **From the south,** directions are from the intersection with Route 28 in Tahoe City.

N4a: Goose Meadows—strolls (6,100') April–August

Driving: **From the north,** drive about 4½ miles south on Route 89, turn left where the sign indicates GOOSE MEADOWS CAMPGROUND and park so as not to block campground traffic. **From the south,** drive about 9 miles north on Route 89 and park on the right as described above.

Walking: Stroll into the meadow.

special gardens: northern suncup, meadow penstemon, Beckwith violet, goldenrod

flowers of note: steershead

N4b: Pole Creek—along road (6,100–7,200'), then if the gate is closed, a 2½-mile hike (if it's open, only about a mile) and difficult scramble up cliffs onto saddle and ridge (if gate is closed, 7,200–8,400'; if gate is open. 7,400–8,400') May–September

Driving: **From the north,** drive about 6 miles south on Route 89, turn right on the Forest Service Road signed POLE CREEK. After about 3 miles, you will reach a gate (closed from Nov. 1-Aug.1) on the left fork. Walk or drive (if the gate is open) up this road. **From the south,** drive about 7 miles north on Route 89, turn left on the Forest Service Road and continue as described above.

Walking: After about 1½ miles from the gate, walk the left fork across the stream. After about ¼ mile, you will come to a clearing on your right across which you can see the volcanic cliffs. Head through the woods to the base of these cliffs and the waterfall/cascade. If you are a strong and careful climber (not technical), you can scramble up about 800' of cliff and contour left (south) and up the draw to the saddle and ridge.

special gardens: porterella, squaw carpet, tiger lily, snowplant, Drummond's anemone, eupatorium, rock fringe, alumroot, fuchsia, alpine paintbrush, Sierra stonecrop, rosy sedum, naked broomrape, checkermallow, horse-mint, mule ears, Copeland's owl's-clover, Anderson's thistle, Lobb's lupine, Torrey's blue-eyed Mary, old man's whiskers, Sierra rein orchid, whorled penstemon

flowers of note: hedgenettle, fringed grass-of-Parnassus, explorer's gentian, Tolmie's saxifrage, Lemmon's keckiella

N4c: Shirley Canyon north—3½-mile hike (6,200–8,400') May–August

Driving: **From the north,** drive about 8¼ miles south on Route 89, turn right on Squaw Valley Road, continue about 2 miles until you reach the 90-degree turn to the left toward the Tram building. Instead of taking this left bend, continue straight into the parking lot. **From the south,** drive about 5 miles north on Route 89 and turn left on Squaw Valley Road and continue as described above.

Walking: Walk to the trailhead just to the right of the Fire Station. After just a few yards, you will intersect the main trail, turn left, and climb all the way to the intersection with the Pacific Crest Trail on the ridge.

special gardens: Leichtlin's mariposa lily, mule ears, scarlet gilia, monkshood, checkermallow, tiger lily, fireweed

flowers of note: skullcap

N4d: Shirley Canyon south—3-mile hike (6,200–8,400') June–September

Driving: Driving directions are the same as for N4c above except take the 90-degree bend to the left. Just before the Tram building, turn right on Squaw Peak Road. Continue on this road as it turns sharply right then left among the condos. Park along the road near the two signs indicating SQUAW PEAK ROAD to the right and SQUAW PEAK WAY to the left.

Walking: Walk up the trail heading north and west along the south side of the creek.

special gardens: Lobb's buckwheat, pretty face

flowers of note: Washington lily, skullcap

N4e: 5-Lakes Basin—4-mile hike (6,400–7,400') June–August

Driving: **From the north,** drive about 10 miles south on Route 89, turn right on Alpine Meadows Road. After about 3 miles park along the right side of the road at the trailhead to 5-Lakes Basin, which is across the road from the 2nd intersection with Deer Park Drive. **From the south,** drive about 3½ miles north on Route 89, turn left on Alpine Meadows Road, and continue as described above.

Walking: Hike up the trail across chapparal slopes and up a draw to the lakes.

special gardens: tobacco brush, mountain violet, Jessica's stickseed

N5: Route 28 from Tahoe City east to Incline Village. From the west, driving directions are from the intersection with Route 89 in Tahoe City. **From the east,** driving directions are from the intersection with Route 267 in Kings Beach.

N5a: Antone Meadows—easy, 2-mile walk (6,700–6,800') June–August

Driving: **From the west,** drive about 2 miles east on Route 28, turn left on Old Mill Road. After about ¼ mile turn left on Polaris Road. After about ½ mile park in the North Tahoe High School lot. **From the east,** drive about 6½ miles west on Route 28, turn right on Old Mill Road, and continue as described above.

Walking: Walk west on the dirt road, very shortly bear right (north) at the T-intersection, walk along the road paralleling the creek to the west for a mile or so. Take the left fork across the creek at your first opportunity, then walk along the west side of the pond and head into the large, open meadow. Head into the woods on the northwest side of the meadow until you find the seep coming down off the rock ledge to the west.

special gardens: Sierra corydalis*

N5b: Dollar Hill—strolls (6,400') June–July

Driving: Driving directions are the same as for N5a above, except park on the shoulder of Route 28 near where Old Mill Road intersects.

Walking: Stroll through the grassy field on the south side of the road.

special gardens: blue flax, wild rose

N6: Route 267 from Truckee south to Kings Beach. From the north, driving directions are from the intersection with Route 80 just east of Truckee. **From the south,** driving directions are from the intersection with Route 28 in Kings Beach.

N6a: Martis Valley—stroll (6,100') April–July

Driving: **From the north,** drive about 3¼ miles south on Route 267, turn right on the dirt road signed MARTIS CREEK WILDLIFE AREA and park at the end. **From the south,** drive about 8¼ miles north on Route 267, turn left into Martis Creek Wildlife Area, and park as described above.

Walking: Wander along the banks below your car and out into the meadows.

special gardens: Beckwith violet, Hooker's balsamroot, Pursh's milkvetch, porterella, northern suncup, star lavender, peony, Anderson's larkspur, water plantain buttercup, camas lily

South of Lake Tahoe

S1: Route 50 from Stateline west to Twin Bridges

S1a: Osgood Swamp—easy, ½-mile walk (6,500') June–October

Driving: From the intersection with Route 89 in South Lake Tahoe, drive about 1¼ miles west on Route 50, park in the gravel turnout on the right side of the road next to the old ski-lift tower.

Walking: Walk down the hill. Just below the gate take the left fork. After a couple hundred yards when the trail bends to the left, take the faint trail to the right into the woods and to the swamp.

special gardens: Sierra rein orchid, monkshood, tiger lily, bog wintergreen, bladderwort*, Brewer's lupine, duck potato

flowers of note: tofieldia, purple cinquefoil, yellow-eyed grass, buckbean, sainfoin

S2: Stateline south and east to Markleeville, Monitor Pass, and Topaz Lake

S2a: Grass Lake—stroll (7,500') June–August

Driving: **From the north,** from the intersection with Route 50 in South Lake Tahoe, drive about 7 miles south on Route 89, park in the gravel turnout on the right on the south side of the lake. **From the south,** from the intersection with Route 88 in Hope Valley, drive about 4 miles north on Route 89, and park as described above.

Walking: Walk through the woods out onto the spongy ground. Go as far as you can (dare) toward the open water of the lake.

special gardens: fireweed, buckbean*, yellow pond lily, mountain violet, azure penstemon

flowers of note: purple cinquefoil, white ladies-tresses, bladderwort, round-leaf sundew

S2b: Freel Peak—strenuous, 5½-mile hike (8,300–10,881') July–September

Driving: **From the north,** from the intersection with Route 50 in South Lake Tahoe, drive about 9 miles south on Route 89 (past the Grass Lake turnout described for S2a above), turn left on the paved road (which will soon turn to dirt and be signed FOREST SERVICE ROAD 051). Stay on 051 for about 3½ miles, just after a bridge over a creek, turn left into a large, dirt parking area. **From the south,** from the intersection with Route 88 in Hope Valley, drive about 2 miles north on Route 89, turn right on the paved road described above.

Walking: Walk up the dirt road (not the road you drove in on) heading west across a short bridge over a creek. After about ½ mile the road turns into a trail; after about another ¼ mile take the right fork signed STARR LAKE—5½ MILES, after about 3½ miles take the right fork at the FREEL PEAK—1 MILE sign.

special gardens: glaucous larkspur, Lewis monkeyflower, alpine penstemon*, alpine gold*, Tolmie's sax-ifrage*, Lobb's buckwheat, alpine paintbrush, monkshood, wallflower, ochre-flowered buckwheat

flowers of note: Sierra podistera

S2c: Hope Valley Wildlife Area—stroll (7,000') May–July

Driving: From the western intersection with Route 88 in Hope Valley below Grass Lake (not the intersection with Route 88 at Woodfords), drive a few yards south into the paved parking area.

Walking: Wander through the meadow.

special gardens: camas lily, California hesperochiron, bog mallow, old man's whiskers, sand corm, northern suncup, bistort

flowers of note: red maids

S2d: Woodfords to Markleeville—strolls along road (5,500') May–July

Driving: From the eastern intersection with Route 88 at Woodfords, drive about 6¼ miles south on Route 89 to Markleeville.

Walking: Explore the gravelly and grassy areas along the road.

special gardens: bird's-foot* lotus, bachelor's buttons*, Torrey's lupine, blepharipappus

S2e: Markleeville south and east over Monitor Pass to Topaz Lake—strolls along road (5,500–8,300–5,100') April–October

Driving: From Markleeville, drive about 5 miles south on Route 89 to the intersection with Route 4, turn left on Route 89, and drive about 17¼ miles east to Topaz Lake.

Walking: Explore the road edges, especially the rock gardens to the east of Monitor Pass and the grassy meadows to the west of the pass.

special gardens: Lemmon's onion, Anderson's larkspur, sand corm, pennyroyal, flat-stemmed onion*, California poppy, water plantain buttercup, aspen onion, desert peach, blue flax, prickly poppy, Wright's buckwheat, checkermallow, horse-mint, showy penstemon

flowers of note: Fremont's phacelia, Bridge's penstemon, narrowleaf milkweed

S3: Route 88 from Woodfords west to Carson Pass area

S3a: Woodfords west to Carson Pass—strolls along road (5,500–8,600') May–October

Driving: From the intersection with Route 89 at Woodfords, drive about 14¼ miles west on Route 88 to Carson Pass.

Walking: Explore the road edges, especially the rocky areas near Carson Pass.

special gardens: tobacco brush, bog mallow, goldenrod, fuchsia, Sierra rein orchid, Torrey's lupine

S3b: Frog Lake and Winnemucca Lake and above—2½-mile hike (8,600–9,500') June–September

Driving: Driving directions are the same as for S3a above, except continue west on Route 88 to just beyond Carson Pass, park on the left in the paved lot (fee area) for the Winnemucca Lake trailhead.

Walking: Hike up the trail about 1 mile to Frog Lake, then across lush seeps to Winnemucca Lake. You might want to scramble up in the rocks above and southwest of the lake.

special gardens: deer's tongue, little elephant's head, broad-leaf lupine, great red paintbrush, red heather, white heather, Sierra primrose, wild iris, bluebells, marsh marigold, old man's whiskers, dwarf Sierra onion, fan-leaf cinquefoil, corn lily, alpine shooting star, alumroot, common yellow monkeyflower, whorled penstemon, mountain pride

flowers of note: alpine fireweed, rosy sedum, alpine buttercup, alpine penstemon

S3c: Meiss Meadows pond trail—easy, 1½-mile hike (8,600–9,000') June–August

Driving: Driving directions are the same as for S3a above, except continue to just beyond the Winnemucca Lake trailhead parking area, park in the paved lot (fee area) on your right at the Meiss Meadows trailhead.

Walking: After about 1½ miles on the trail, you will reach a pond and large meadow.

special gardens: deer's tongue, old man's whiskers, bluebells, wild iris, fan-leaf cinquefoil, Copeland's owl's-clover, narrow-leafed stonecrop, Leichtlin's mariposa lily

S3d: Red Lake Peak summit—2½ mile hike (8,600–10,000') June–August

Driving: Driving and parking directions are the same as for S3c above.

Walking: Hike the same trail as for S3c above, except that at the pond, head cross-country east up to the summit of Red Lake Peak.

special gardens: Lobb's lupine, pink gilia, deer's tongue, bluebells, primrose monkeyflower, little elephant's head, low cryptantha*, dwarf Sierra onion, Pursh's milkvetch, alpine paintbrush, Shasta clover, daggerpod

West of Lake Tahoe

W1: Tahoe City south to Stateline. From the north, driving directions are from the intersection with Route 28 in Tahoe City. **From the south,** driving directions are from the intersection with Route 50 in Stateline.

W1a: Sherwood Forest—1½-mile hike and scramble (7,100–8,300') June–August

Driving: **From the north,** drive about 2¼ miles south on Route 89, turn right on Pineland Drive (two columns say PINELAND), take the left fork on Twin Peak Road (turns into Ward Creek Boulevard). After about 4¼ miles at road end (it loops back), park on the road edge. **From the south,** drive about 24¼ miles north on Route 89, turn left on Pineland Drive, and continue as described above.

Walking: Hike cross-country west across the field and scramble up to the ridge directly to the east. You will have to cross a willow-choked creek in a narrow canyon.

special gardens: rock fringe, flat-stemmed onion*, great red paintbrush, little elephant's head

flowers of note: rosy sedum, dwarf hesperochiron

W1b: Paige Meadows—easy, 2-mile walk (6,400–7,000') June–August

Driving: Driving directions are the same as for W1a above, except that after about 2 miles from Route 89 on Pineland Drive and Ward Creek Boulevard, park on the gravel shoulder on the left side of the road, directly opposite the gated jeep road to the right heading up the hill to Paige Meadows.

Walking: Head up the jeep road; at your first intersection (a branch to the right), stay left. About a mile from the start, when the road bends right, walk straight ahead off the road into the large, wet meadow. Explore carefully, as it's easy to get disoriented in this string of meadows.

special gardens: monkshood, bull elephant's head, bog mallow, horse-mint, porterella, camas lily, broad-leaf lupine, Sierra rein orchid, crimson columbine, corn lily, bistort

W1c: Blackwood Canyon to Barker Pass—strolls along the road (6,300–7,500') June–August

Driving: **From the north,** drive about 4 miles south on Route 89, turn right on Blackwood Canyon Road (Sno-Park area). Continue several miles until shortly after the road turns to dirt, turn right into a parking area in the woods at a trail sign for Pacific Crest Trail. **From the south,** drive about 22½ miles north on Route 89, turn left on Blackwood Canyon Road, and continue as described above.

Walking: Explore the road edges, especially the seep areas as the road heads up toward Barker Pass.

special gardens: Lewis monkeyflower, Torrey's lupine

flowers of note: buck lotus

W1d: Barker Peak—2-mile walk and scramble up rocks to the summit (7,500–8,166') June–August

Driving: Driving and parking directions are the same as for W1c above.

Walking: Walk north from the parking area up the trail signed PCT TO TWIN PEAKS. After you cross the small creek and contour the slope, you will reach a saddle. Cross-country south up the ridge and talus slope to the summit. The last stretch is over some large, sometimes-loose rocks, so be careful.

special gardens: grand collomia, mule ears, Copeland's owl's-clover, pennyroyal

W1e: Barker Pass south—easy, ½-mile walk (7,500–7,000') June–August

Driving: Driving and parking directions are the same as for W1c above.

Walking: From the parking area, walk back up to the dirt road, head left, then within a few yards take the PCT south into the large seep areas.

special gardens: scarlet gilia, Anderson's thistle, glaucous larkspur, broad-leaf lupine, Torrey's blue-eyed Mary

flowers of note: naked star tulip, bowl clover

W1f: Meeks Bay—3-mile hike (6,200–7,000') May–August

Driving: **From the north,** drive about 11 miles south on Route 89, park in the paved lot on the right for the Desolation Wilderness trailhead (self-registration required, no fee). **From the south,** drive about 16 miles north on Route 89, park on the left as described above.

Walking: Walk up the level trail/jeep road. At the fork take the right branch, which climbs along the creek.

flowers of note: Sierra lupine, Tahoe yellow cress (on Lake Tahoe beach)

W1g: Vikingsholm—1-mile walk (6,600–6,200') May–July

Driving: **From the north,** drive about 18 miles south on Route 89, turn into the paved parking lot on the left signed EMERALD BAY STATE PARK VIKINGSHOM. **From the south,** drive about 8½ miles north on Route 89 and park on the right as described above.

Walking: Walk down the steep, paved trail to Vikingsholm You might want to walk the lakeside trail to Emerald Point at the bottom (to see the rare saprophytic sugarstick).

special gardens: alumroot, greenleaf manzanita, Lobb's nama*, mountain pride

flowers of note: sugarstick, purple nightshade, yellow-eyed grass

W1h: Eagle Lake—steep, 1-mile hike (6,500–7,200') June–August

Driving: Driving directions are the same as for W1g above, except park in the paved lot for the Eagle Falls trailhead a few hundred yards south of the Vikingsholm parking lot.

Walking: Head up the steep trail to Eagle Lake.

special gardens: creek dogwood, alumroot

W1i: Mt. Tallac—strenuous, 4½-mile hike (6,400–9,735') June–September

Driving: **From the north,** drive about 23 miles south on Route 89, turn right at Camp Shelley/Camp Concord. After about ¼ mile, turn left at the Tallac trailhead sign and continue about .6 mile to the trailhead. **From the south,** drive about 4 miles north on Route 89, turn left at Camp Shelley/Camp Concord/Mt. Tallac and continue as described above.

Walking: Hike the steep trail past Floating Island Lake and Cathedral Lake across the talus slopes and meadows to the summit.

special gardens: Sierra stonecrop, seep-spring arnica, little elephant's head, fireweed, red heather, bog wintergreen, eupatorium, wallflower, snowberry, Leichtlin's mariposa lily, rock fringe, Lewis monkeyflower, white heather, alpine gold*, pussypaws

flowers of note: explorer's gentian, showy polemonium, alpine gentian, Washington lily

W1j: Taylor Creek-strolls along the road (6,300–6,500') May–June

Driving: **From the north,** drive about 23 miles south on Route 89, turn right at the first road after the Camp Shelley/Camp Concord road. After about ¹⁄₁₀ mile turn right just before the Sno-Park sign and continue about 2 miles to the end of public access. **From the south,** drive about 3½ miles north on Route 89, turn left after the Sno-Park sign and continue as described above.

Walking: Explore the wet meadows along the road.

special gardens: camas lily, bistort

FEBRUARY

In most years, the Tahoe Basin is deep in snow in February with frequent storms adding to the total. In years of heavy and/or late snow and a cold spring, you will probably have to wait until well into March for the first blooming down in Carson City. In most years of average or below-average snow and a warm spring, a short drive from the Tahoe Basin to the Great Basin desert to the east (especially Carson Valley to the southeast) will reward you with some green stirrings (sprouting and leafing) and even some actual flowers—only the bare beginnings of the flower season, of course, but what a joy it is!

If it's an early blooming year, it could well be worth your while in late February to head down to some of the vacant fields in and around Carson City. Even in the years of lightest winter, you won't find much in bloom, but you are likely to find a few dashes of mild color. You will have to look carefully, as these are small plants with tiny flowers close to the ground that tend to blend in with their rocky or sandy habitats, but when you find them, you will feel a surge of spring—before you head back up to winter in the Tahoe Basin.

BUR BUTTERCUP
Ranunculus testiculatus
Buttercup Family (Ranunculaceae)

Common on dry flats in the Great Basin at low elevation. Short plant; many 5-petaled flowers; deeply divided leaves with fingerlike lobes; cluster of spiked beaks in seed. Alien.

Carson City (E1g), 4,700', February–April

Peavine Mt. east (E1a), 5,000', March–May

NEVADA LOMATIUM
Lomatium nevadense
Carrot Family (Apiaceae)

Occasional on dry slopes in eastern and western Tahoe and in the Great Basin at low and mid elevation. Plant nearly flat on ground; downy leaves; often blooms between snow patches.

Carson City (E1g), 4,700', February–March

Thomas Creek lower trailhead (E4f), 6,000', March–April

Sherwood Forest (W1a), 8,000', June

SPRING WHITLOWGRASS
Draba verna
Mustard Family (Brassicaceae)

Common in grassy fields in the Great Basin at low elevation. Short plant with tiny flowers; 4 deeply 2-lobed petals.

Carson City (E1g), 4,700', February–March

Topaz Lake (E1j), 5,100', March–April

MARCH

Although most of the Tahoe Basin is still covered with snow in March, in most years many south-facing slopes are beginning to show bare ground with definite signs of plant growth. This is a wonderful month to see wildflowers if you can drive a few hours to the west and southwest to the Central Valley or to the southeast to the Mojave Desert and Death Valley. If you're lucky enough to get to Death Valley this month in a year of better-than-average winter and spring rain, you will be dazzled and amazed with the explosion of annuals covering the usually bare alluvial fans with solid masses of blooms.

In all but the latest blooming years, you can also find wonderful spring blooming in March with only a short drive to the east of the Tahoe Basin. Although not as spectacular as the Central Valley or Death Valley, the sagebrush flats in Carson City, along Deadman's Creek to the east of Washoe Lake, and along the west side of Topaz Lake will be coming alive with wildflower color. In all but those years of "everlasting" winter, by the latter part of March, you are likely to find great gardens, especially in the Carson City fields, where the earliest blooming flowers of February and early March are now joined by many other species, some with much larger flowers, to form wonderful palettes of yellow, blue-purple, pink, rose, and white.

SILVER LUPINE
Lupinus argenteus
Pea Family (Fabaceae)

Common on dry flats and slopes in eastern and southern Tahoe and in the Great Basin at low and mid elevation. Plant covered with satiny, silver hairs; racemes of flowers rise well above leaves.

Carson City (E1g), 4,700', March–May

Topaz Lake (E1j), 5,100', April–May

Peavine Mt. east (E1a), 5,500', June

Mt. Rose (E4c), 9,000', June–August

Galena Creek (E4e), 7,500', July

Freel Peak trail (S2b), 8,500', July–August

WASHOE PHACELIA
Phacelia curvipes
Waterleaf Family (Hydrophyllaceae)

Rare, on dry flats and slopes in the Great Basin at low elevation. Short stems with few violet-colored flowers with white throats; hairy plant.

Deadman's Creek (E1f), 5,000', March–May

ANDERSON'S MILKVETCH
Astragalus andersonii
Pea Family (Fabaceae)

Occasional on dry, sandy flats in the Great Basin at low elevation. Pale pink or yellowish flowers with pink or lavender tinges; white-hairy plant; seedpods are roundish, stiffly papery, and hairy.

Deadman's Creek (E1f), 5,000', March–May

Topaz Lake (E1j), 5,100', May

Carson City (E1g), 4,700', May

Ophir Creek trail (E1d), 5,600', May–June

Peavine Mt. east (E1a), 6,000', May–June

DWARF ONION
Allium parvum
Lily Family (Liliaceae)

Occasional on dry, sandy, or gravelly flats in the Great Basin at low and mid elevation. Flowers nearly on the ground; 2 prostrate leaves that extend out beyond the flowers; tepals range from deep rose to pale pink to white.

Carson City (E1g), 4,700', March–April

Thomas Creek lower trailhead (E4f), 6,000', April–May

Old Geiger Grade (E1c), 6,200', April–May

LONG-LEAF PHLOX
Phlox longifolia
Phlox Family (Polemoniaceae)

Common on dry slopes in the Great Basin at low and mid elevation. Short plant; narrow but not needle-like, hairy leaves; narrow, clawed petals range from rose to pink to white.

Carson City (E1g), 4,700', March–April

Topaz Lake (E1j), 5,100', March–April

Thomas Creek lower trailhead (E4f), 6,000', April–May

Peavine Mt. south (E1b), 7,000', April–May

Old Geiger Grade (E1c), 6,000', April–May

DESERT PEACH
Prunus andersonii
Rose Family (Rosaceae)

Common on dry flats and slopes in southeastern Tahoe and in the Great Basin at low and mid elevation. Shrub thick with large pink or rose flowers; small, peachlike fruit.

Washoe Valley (E1e), 4,700', March–April

Carson City (E1g), 4,700', March–April

Deadman's Creek (E1f), 5,100', March–May

Faye-Luther trail (E1i), 5,500', April

east of Markleeville (S2e), 5,800', April–May

west of Woodfords (S3a), 6,000', April–May

BECKWITH VIOLET
Viola beckwithii
Violet Family (Violaceae)

Occasional (locally common) on dry, sandy flats in northern Tahoe and in the Great Basin at low and mid elevation. Large flowers rising a few inches above basal leaves; upper 2 petals velvety purple-maroon, lower 3 petals blue-purple, maroon, pink, or white; fan-shaped leaves divided into linear segments.

Peavine Mt. east (E1a), 5,500', March–April

Thomas Creek lower trailhead (E4f), 6,500', March–May

Old Geiger Grade (E1c), 6,000', April–May

Martis Valley (N6a), 6,100', April–May

Goose Meadows (N4a), 6,100', April–May

HOOKER'S BALSAMROOT
Balsamorhiza hookeri
Sunflower Family (Asteraceae)

Occasional on dry, sagebrush slopes in northern Tahoe and in the Great Basin at low elevation. Large flower heads; long, narrow, deeply pinnately divided leaves.

Carson City (E1g), 4,700', March–April

Martis Valley (N6a), 6,100', April–May

Deadman's Creek (E1f), 5,100', May

Peavine Mt. east (E1a), 5,500', May

Old Geiger Grade (E1c), 6,000'. April–June

DESERT GOOSEBERRY

Ribes velutinum
Gooseberry Family (Grossulariaceae)

Occasional on dry slopes, canyons, washes in the Great Basin at low and mid elevation. Shrub with 1–3 nodal spines; short, tubular, pale-yellow, pink, or white flowers; small, 3–5 lobed leaves; round, velvety hairy, red-purple berry.

Deadman's Creek (E1f), 5,000', March–April

VEATCH'S STICKLEAF

Mentzelia veatchiana
Loasa Family (Loasaceae)

Common on dry slopes and flats in the Great Basin at low and mid elevation. White-hairy leaves deeply cleft into narrow lobes; satiny yellow petals deep orange at the base (orange "eye" on corolla).

Deadman's Creek (E1f), 5,000', March–April

Peavine Mt. south (E1b), 5,000', April–May

TINY EVENING-PRIMROSE

Camissonia pusilla
Evening-Primrose Family (Onagraceae)

Occasional on dry flats and slopes in eastern Tahoe and in the Great Basin at low and mid elevation. Short plant with wiry, often red, stems; small, 4-petaled flower; globular stigma; red buds.

Carson City (E1g), 4,700', March–April

Thomas Creek lower trailhead (E4f), 6,000', April–May

Ophir Creek trail (E1d), 5,300', April–May

Peavine Mt. south (E1b), 5,500', May

above Marlette Lake (E2c), 8,000', July

YELLOW-OR-PURPLE MONKEYFLOWER (YELLOW)
Mimulus densus
Figwort Family (Scrophulariaceae)

Occasional on dry flats in the Great Basin at low elevation. Dwarfed plant with large flowers; red splotches on petals; glandular calyx and leaves, which have strong, pungent odor when crushed.

Carson City (E1g), 4,700', March–April

WHITE LAYIA
Layia glandulosa
Sunflower Family (Asteraceae)

Occasional on sandy flats in the Great Basin at low elevation. Large flowerheads with 3-lobed, rounded rays; large, raised, yellow disk; sticky, fragrant stem and leaves; often grows in masses.

Carson City (E1g), 4,700', March–April

Washoe Valley (E1e), 5,000', March–May

Peavine Mt. south (E1b), 5,000', April–May

Old Geiger Grade (E1c), 5,000', April–May

Ophir Creek (E1d), 5,500', April–May

SLENDER POPCORN FLOWER
Plagiobothrys tenellus
Borage Family (Boraginaceae)

Common on dry flats and slopes in the Great Basin at low elevation. Short, hairy plant; basal rosette of leaves with a few scattered stem leaves; tiny, creamy white flowers with a yellow "eye."

Carson City (E1g), 4,700', March–April

Peavine Mt. east (E1a), 6,000', April–May

APRIL

Most years in April the bare patches of March have grown to large expanses of bare ground at lake level, though snow will still be quite widespread and thick at higher elevations. At lake level, however, life is definitely stirring—stems and leaves are apparent all over, and buds are beginning to strain. In some of the lowest elevation meadows north of the lake, you can find a few species that are the first colorful ripple of the wildflower tide soon to come. Most of these species usually grow in large patches, creating stunning canvases.

If you want to see more diverse gardens, head east of the Tahoe Basin, where you are likely to find gorgeous multihued gardens of color and fragrance. The gardens of the "flatlands" on the floor of Carson Valley and Washoe Valley are creeping up into the low hills—the bases of the Virginia Range to the east and of the Carson Range to the west are coming alive with flowers.

Particularly lush, dense, and varied gardens can be found within a couple hundred yards of the Thomas Creek lower trailhead. If you want to see fields of orange, pink, and purple, head to the lower elevations of the south side of Peavine Mountain. And if you want to revel in the colors of spring against a backdrop of winter, go to Jack's Valley Habitat Management Area.

Drive up the first few miles of Route 89 west of Topaz Lake toward Monitor Pass to see the yellow-throated, pale purple, tubular flowers of Fremont's phacelia, otherwise unknown in the Tahoe area.

STAR LAVENDER/CAT'S BREECHES
Hydrophyllum capitatum
Waterleaf Family (Hydrophyllaceae)

Common on dry flats (usually with sagebrush) throughout Tahoe and in the Great Basin at low and mid elevation. Round clusters of fuzzy, pale blue flowers nearly on the ground under a canopy of deeply divided leaves.

Thomas Creek lower trailhead (E4f), 6,000', April–May

Peavine Mt. east (E1a), 5,500', April–May

Sagehen Creek west (N1b), 6,200', May–June

west of Monitor Pass (S2e), 7,000', May–June

Martis Valley (N6a), 6,100', May–June

FREMONT'S PHACELIA
Phacelia fremontii
Waterleaf Family (Hydrophyllaceae)

Rare, on dry slopes in the Great Basin at low elevation. Large, blue-purple or pink-purple flowers with yellow throats; white-hairy, pinnately lobed leaves with round lobes.

west of Topaz Lake (S2e), 5,300', April–May

COMMON PURPLE MUSTARD
Chorispora tenella
Mustard Family (Brassicaceae)

Common on dry flats, disturbed places in the Great Basin at low and mid elevation. Tall plant; many small, purple flowers branch off stem; alternating, fleshy, tonguelike leaves; long, narrow, cylindrical, upcurving seedpods. Alien.

Old Geiger Grade (E1c), 5,000', April–May

Deadman's Creek (E1f), 5,000', April–May

Peavine Mt. south (E1b), 6,500', April–May

DAGGERPOD
Phoenicaulis cheiranthoides
Mustard Family (Brassicaceae)

Occasional on dry, rocky slopes and ridges throughout Tahoe and in the Great Basin at low, mid, and high elevation below timberline. Basal cluster of furry, tonguelike leaves; cylindrical racemes of 4-petaled, red-purple flowers; seedpods are sharp-edged "blades."

Deadman's Creek (E1f), 5,200', April–May

Peavine Mt. east (E1a), 5,500', April–May

Mt. Rose (E4c), 10,000', June–July

Red Lake Peak (S3d), 9,500', June–July

Pole Creek headwaters (N4b), 7,700', June–July

above Winnemucca Lake (S3b), 9,500', June–July

SHAGGY MILKVETCH
Astragalus malacus
Pea Family (Fabaceae)

Occasional on dry flats and slopes in the Great Basin at low elevation. Densely white-hairy plant; dark red-purple veins on banner; shaggy, slightly upcurved seedpods,

Carson City (E1g), 4,700', April–May

Peavine Mt. south (E1b), 5,000', April–May

Deadman's Creek (E1f), 5,200', April–May

PURSH'S MILKVETCH
Astragalus purshii
Pea Family (Fabaceae)

Occasional on dry flats and slopes throughout Tahoe and in the Great Basin at low, mid, and high elevation below timberline. Prostrate plant covered with soft, cottony hairs; rounded leaflets; pink-purple flowers with large, white patch on the banner; roundish "cotton ball" seedpods.

Old Geiger Grade (E1c), 6,100', April–May

Thomas Creek lower trailhead (E4f), 6,000', April–May

Peavine Mt. east (E1a), 6,000', April–May

Martis Valley (N6a), 6,100', May–June

Thomas Creek upper trail (E4g), 7,000', June

Red Lake Peak (S3d), 9,800', June

ANDERSON'S CLOVER
Trifolium andersonii
Pea Family (Fabaceae)

Occasional on dry flats and slopes in the Great Basin at low elevation. Mat of leaves with usually 5 leaflets; spherical clusters of rose-pink flowers; plant densely silky-hairy.

Peavine Mt. east (E1a), 5,500', April–May

Old Geiger Grade (E1c), 6,200', April–May

BIG-HEADED CLOVER
Trifolium macrocephalum
Pea Family (Fabaceae)

Occasional (locally common) on dry flats and slopes in the Great Basin at low elevation. Sprawling plant; palmately compound leaves with 5–9 toothed leaflets; large, roundish heads of rose, pink, or white flowers; often grows in masses.

Thomas Creek lower trailhead (E4f), 6,000', April–May

Peavine Mt. east (E1a), 5,500', April–May

Peavine Mt. south (E1b), 5,500', April–May

KELLOGG'S ONION
Allium anceps
Lily Family (Liliaceae)

Rare; on dry flats and slopes in the Great Basin at low elevation. Short plant; very narrow (almost threadlike), widely separated, pink or rose-purple tepals; carnation-like fragrance.

Geiger Grade (E1c), 5,000', April–May

Peavine Mt. east (E1a), 5,500', April–May

DESERT PAINTBRUSH
Castilleja chromosa
Figwort Family (Scrophulariaceae)

Occasional on dry, sandy flats and slopes in the Great Basin at low and mid elevation. Scarlet bracts (sometimes orange or yellow) dark red-brown at base; bracts and leaves with 3 or 5 lobes.

Old Geiger Grade (E1c), 5,000', April–May

Deadman's Creek (E1f), 5,000', April–May

Thomas Creek lower trailhead (E4f), 6,000', April–May

Jack's Valley Habitat Management Area (E1h), 5,100', May

Peavine Mt. south (E1b), 6,500', May–June

SONNE'S DESERT PARSLEY

Lomatium austinae/L. plummerae var. *sonnei*
Carrot Family (Apiaceae)

Common on dry sagebrush slopes in the Great Basin at low elevation. Large flower umbels on usually prostrate stems; blue-green, fernlike leaves covered with downy, white hairs.

Peavine Mt. east (E1a), 5,500', April

Old Geiger Grade (E1c), 6,200', April

Thomas Creek lower trailhead (E4f), 6,000', April–May

ARROWLEAF BALSAMROOT

Balsamorhiza sagittata
Sunflower Family (Asteraceae)

Common on dry, sandy slopes throughout Tahoe and in the Great Basin at low and mid elevation. Large flower heads; large, broad, shiny, arrow-shaped leaves on arching petioles.

Deadman's Creek (E1f), 5,100', April–May

Jack's Valley Habitat Mangagement Area (E1h), 5,100', April–May

Shirley Canyon north (N4c), 6,200', June

Paige Meadows (W1b), 6,400', June

Peavine Mt. east (E1a), 7,500', June

SINGLE-STEM SENECIO/ SINGLE-STEM GROUNDSEL

Senecio integerrimus
Sunflower Family (Asteraceae)

Common in dry or moist fields, forest openings throughout Tahoe and in the Great Basin at low and mid elevation. Single, stout stem; long-petioled, round or tonguelike basal leaves; flower heads with few irregularly spaced rays.

Thomas Creek lower trailhead (E4f), 6,000', April–May

Deadman's Creek (E1f), 5,100', April–May

Kyburz Flat (N1a), 5,900', April–June

Peavine Mt. east (E1a), 6,000', April–June

Meeks Bay (W1f), 6,300', May–June

Winnemucca Lake trail (S3b), 8,700', June

FIDDLENECK
Amsinckia tessellata
Borage Family (Boraginaceae)

Occasional (locally common) on dry, sandy slopes in the Great Basin at low elevation. Plant bristly with stiff, white hairs that can cause a rash; long, curling, caterpillar-like spikes of tubular, yellow-orange flowers.

Carson City (E1g), 4,700', April–May

Old Geiger Grade (E1c), 5,000', April–May

Peavine Mt. south (E1b), 5,000', April–May

GOLDEN CURRANT
Ribes aureum
Gooseberry Family (Grossulariaceae)

Occasional in moist places on sagebrush flats and slopes, canyons in the Great Basin at low elevation. Shrub with smooth, 3-lobed leaves with spicy fragrance; golden yellow tubular flowers.

Peavine Mt. south (E1b), 5,500', April–May

YELLOW CLUB-FRUITED EVENING-PRIMROSE
Camissonia claviformis
Evening-Primrose Family (Onagraceae)

Occasional on dry flats and slopes in the Great Basin at low elevation. Flowers in clusters toward tip of stems; 4 petals with red spots at base of petals; long-petioled basal leaves with red spots; several small pinnate leaflets and one large, blunt, terminal leaflet.

Steamboat Springs (E1e), 4,600', April–May

Peavine Mt. south (E1b), 5,000', April–May

Old Geiger Grade (E1c), 5,000', May

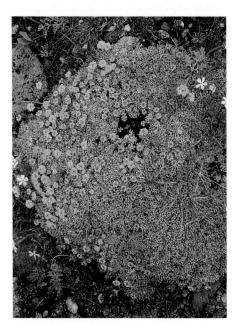

CUSHION DESERT BUCKWHEAT
Eriogonum caespitosum
Buckwheat Family (Polygonaceae)

Common on dry slopes in the Great Basin at low elevation. Short stems rise above tight mat or cushion of felty leaves; white-hairy bracts under flower heads; bright yellow flowers turn deep red with age.

Thomas Creek lower trailhead (E4f), 6,000', April–May

Old Geiger Grade (E1c), 6,200', April–May

Peavine Mt. east (E1a), 5,500', April–May

WATER-PLANTAIN BUTTERCUP
Ranunculus alismifolius
Buttercup Family (Ranunculaceae)

Common in wet meadows and marshes and along creeks throughout Tahoe at low and mid elevation. Narrow, fleshy, unlobed leaves; variable number of shiny petals; grows in masses.

Sagehen Creek east (N1c), 5,900', April–May

Martis Valley (N6a), 6,100', April–May

west of Monitor Pass (S2e), 7,000', May

Pole Creek (N4b), 6,700', June

Tahoe Meadows (E4b), 8,600', June

Winnemucca Lake trail (S3b), 8,600', June

DESERT BUTTERCUP
Ranunculus cymbalaria
Buttercup Family (Ranunculaceae)

Occasional in wet or moist areas among sagebrush in the Great Basin at low and mid elevation. Large, round, unlobed or 3-lobed leaves on long petioles; 5–12 large, shiny petals; 5 yellow-green sepals.

Thomas Creek lower trailhead (E4f), 6,000', April

Peavine Mt. south (E1b), 5,000', April

BITTERBRUSH
Purshia tridentata
Rose Family (Rosaceae)

Common on dry slopes throughout Tahoe and in the Great Basin at low and mid elevation. Shrub that often covers hillsides; cinnamon-like fragrance; 3-lobed, mittenlike leaves.

Jack's Valley Habitat Management Area (E1h), 5,000', April–May

Deadman's Creek (E1f), 5,000', April–May

Ophir Creek trail (E1d), 5,200', May

Thomas Creek upper trail (E4g), 7,000', June–July

Mt. Rose (E4c), 9,000', June–July

Frog Lake (S3b), 8,800', June–July

YELLOW VIOLET
Viola aurea
Violet Family (Violacea)

Occasional on dry flats and slopes in the Great Basin at low and mid elevation. Low plant; loosely hairy stems and leaves; round or ovate, gray-green leaves; purple veins on lower, middle petal; back of upper two petals brown-purple.

Thomas Creek lower trailhead (E4f), 6,000', April–May

Peavine Mt. south (E1b), 5,000', April–May

SAND CORM
Zigadenus paniculatus
Lily Family (Liliaceae)

Common in dry areas, usually with sagebrush, throughout Tahoe and in the Great Basin at low and mid elevation. 6 small tepals with yellow-green splotch at base; showy, yellow anthers; long, grasslike leaves. Poisonous.

Thomas Creek lower trailhead (E4f), 6,000', April–May

Carson City (E1g), 4,700', April–May

Peavine Mt. east (E1a), 5,500', May–June

east of Monitor Pass (S2e), 5,300', May–June

Hope Valley Wildlife Area (S2c), 7,000', June

STEERSHEAD
Dicentra uniflora
Poppy Family (Papaveraceae)

Occasional on sandy or gravelly flats in northern, western, and southern Tahoe at low and mid elevation. White or pink flowers resemble bleached steer skull; single flower on short stem; divided leaf on separate stalk.

Goose Meadows (N4a), 6,100', April–May
Donner Pass Road (N3a), 6,800', June
Castle Peak east (N2b), 8,800', June

SPREADING PHLOX
Phlox diffusa
Phlox Family (Polemoniaceae)

Common on dry, rocky flats and slopes throughout Tahoe and in the Great Basin at low, mid, and high elevation to above timberline. White or pink pinwheel flowers sit atop loose mat of prickly, needle-like leaves

Deadman's Creek (E1f), 5,200', April–May

Shirley Canyon north (N4c), 6,200', May–June

Mt. Rose (E4c), 10,700', June–July

Tamarack Lake (E4d), 8,700', June–July

above Winnemucca Lake (S3b), 9,400', June–July

Mt. Tallac (W1i), 9,500', July

MARSH MARIGOLD
Caltha leptosepala
Buttercup Family (Ranunculaceae)

Common in wet meadows and bogs throughout Tahoe at low, mid, and high elevations below timberline. Large, bright flowers with variable number of petal-like sepals; kidney-shaped, glossy leaves.

Sagehen Creek west (N1b), 6,200', April–June

Mt. Rose (E4c), 9,000', June

Tahoe Meadows (E4b), 8,700', June

above Winnemucca Lake (S3b), 9,200', June–July

SLENDER WOODLAND STAR
Lithophragma glabrum
Saxifrage Family (Saxifragaceae)

Occasional in open woods, damp flats, and slopes throughout Tahoe and in the Great Basin at low and mid elevation. Delicate white or pink petals with 4 or 5 unequal, feathery lobes; leaves cut into many lobes; often red, hairy, bb-like "bulbils" in upper leaf axils.

Thomas Creek lower trailhead (E4f), 6,000', April–May

Peavine Mt. east (E1a), 5,500', April–May

Pole Creek headwaters (N4b), 7,600', June–July

Mt. Tallac (W1i), 7,500', June–July

WESTERN PEONY
Paeonia brownii
Peony Family (Paeoniaceae)

Occasional on dry flats, forest openings throughout Tahoe and in the Great Basin at low and mid elevation. Large, nodding, maroon flowers; deeply lobed leaves; very large, sausage-like seed pods.

Peavine Mt. east (E1a), 5,500', April–May

Martis Valley (N6a), 6,100', May

Sagehen Creek west (N1b), 6,200', May

Spooner Lake (E2b), 7,100', May–June

MAY

By May in most years, the gardens in Washoe Valley and Carson Valley and in the foothills of the Carson Range and the Virginia Range are peaking in glorious "summer" bloom. For midseason wildflower bliss be sure to visit high desert "hot spots" to the east of the Tahoe Basin. Climb up the first mile or so of the trails and roads ascending the eastern escarpment of the Carson Range (to about 7,500') to find dazzling displays.

Significant and spectacular blooming has also come to Tahoe in May, mostly in the low-lying meadows north of the lake but also in some higher-elevation meadows to the south. The sunny meadows north and south of Lake Tahoe are spectacular appetizers for the main course to come. Although you will probably find beginnings of blooming in the lowest parts of trails climbing from the west shore of Lake Tahoe up toward the crest of the Sierra, this west shore gets heavy winter snow so only the earliest spring bloomers are likely to be found.

In mid-to-late May, be sure to visit the incredible acres of camas lily just west of Stampede Reservoir along Sagehen Creek East—the most spectacular "lake" of these gorgeous, dark blue-purple lilies anywhere in the Tahoe area and maybe anywhere in the Sierra. For gorgeous May meadows south of the Lake, from Topaz Lake head west up and over Monitor Pass on Route 89. On the east side of the pass, you'll find great rock gardens with the uncommon flat-stemmed onion, and on the west side of the pass, you'll be dazzled by Monet paintings of gold, blue-purple, and rose.

BACHELOR'S BUTTONS
Centaurea cyanus
Sunflower Family (Asteraceae)

Occasional on dry, sandy flats, disturbed places in eastern Tahoe and in the Great Basin at low elevation. Large flower heads with enlarged disk flowers that appear to be flaring rays; green phyllaries tipped with tiny, white or black teeth; ribbed stem. Alien.

Woodfords (S2d), 5,500', May–July

TORREY'S LUPINE
Lupinus lepidus var. *sellulus*
Pea Family (Fabaceae)

Common on dry, rocky flats and slopes, forest openings, roadsides, and other disturbed places throughout Tahoe and in the Great Basin at low and mid elevation. Short plant; very dense, neatly cylindrical racemes rising cleanly above basal leaves; flowers tend toward violet.

Washoe Valley (E1e), 4,600', May

Blackwood Canyon (W1c), 6,300', June

Woodfords to Markleeville (S2d), 5,500', June

Woodfords to Carson Pass (S3a), 7,200', June–July

Spooner Lake (E2b), 7,000', June–July

DWARF PHACELIA
Phacelia humilis
Waterleaf Family (Hydrophyllaceae)

Common on dry flats, forest openings in northern and eastern Tahoe and in the Great Basin at low and mid elevation. Small, bowl-shaped flowers; plants usually form large carpets; tonguelike, deeply veined leaves.

Kyburz Flat (N1a), 5,900', May

Deadman's Creek (E1f), 5,200', May

Faye-Luther trail (E1i), 6,000', May–June

Sagehen Creek west (N1b), 6,200', June

Galena Creek (E4e), 8,000', June–July

Tamarack Lake (E4d), 8,700', July

NARROW-LEAF PHACELIA
Phacelia linearis
Waterleaf Family (Hydrophyllaceae)

Occasional on dry, sandy, or rocky flats and slopes, usually with sagebrush, in the Great Basin at low and mid elevation. Violet flowers in small clusters (not caterpillar-like cymes); narrow, unlobed leaves, except for a few short, needle-like segments.

Faye-Luther trail (E1i), 4.900', May

Peavine Mt. east (E1a), 5,500', June

WILD IRIS
Iris missouriensis
Iris Family (Iridaceae)

Occasional in wet meadows, seeps, disturbed places in eastern and southern Tahoe and in the Great Basin at low and mid elevation. Tall plant; large, showy flowers with 3 blue-purple, tonguelike sepals with networks of white and yellow veins; 3 narrow, erect petals; grasslike leaves.

Gardnerville to Woodfords (E3a), 4,800', May–June

Washoe Valley (E1e), 5,000', May–June

Thomas Creek upper trail (E4g), 7,500', June

Frog Lake (S3b), 8,900', June–July

Meiss Meadows pond (S3c), 9,000', June–July

CAMAS LILY
Camassia quamash
Lily Family (Liliaceae)

Common in wet, grassy meadows, streambanks throughout Tahoe at low and mid elevation. Star-shaped flowers with 6 deep blue tepals; long, grasslike leaves; grows in masses (sometimes acres).

Kyburz Flat (N1a), 5,900', May

Sagehen Creek east (N1c), 5,900', May

Martis Valley (N6a), 6,100', May

Taylor Creek (W1j), 6,300', May

Hope Valley Wildlife Area (S2c), 7,000', May–June

Paige Meadows (W1b), 7,000', June

BLUE FLAX
Linum lewisii
Flax Family (Linaceae)

Occasional on dry flats and slopes throughout Tahoe at low, mid, and high elevation below timberline. Tall plant with slender, sinewy stems; fragile, steely blue petals that fall off easily; often grows in masses.

east of Monitor Pass (S2e), 6,200', May–June

Dollar Hill (N5b), 6,400', June–July

Mt. Rose (E4c), 10,200', June–July

Sherwood Forest (W1a), 7,500', June–July

Shirley Canyon north (N4d), 6,300', June–July

Red Lake Peak (S3d), 9,800', July

ANDERSON'S LARKSPUR
Delphinium andersonii
Buttercup Family (Ranunculaceae)

Common on dry flats and slopes in northern and eastern Tahoe and in the Great Basin at low and mid elevation. Loosely flowered; upper 2 petals white with blue tinge; stout, smooth, reddish stem; deeply palmately lobed leaves.

Thomas Creek lower trailhead (E4f), 6,100', May–June

Martis Valley (N6a), 6,100', May–June

east of Monitor Pass (S2e), 5,300', May–June

Thomas Creek upper trail (E4g), 7,600', June–July

Peavine Mt. east (E1a), 7,500', June–July

SQUAW CARPET/MAHALA MAT
Ceanothus prostratus
Buckthorn Family (Rhamnaceae)

Common on dry slopes, forest openings in northern, western, and eastern Tahoe at low and mid elevation. Clusters of small, pale blue flowers with clawed petals and grapelike fragrance; stiff, holly-like leaves; grows in large patches or masses that contour ground.

Ophir Creek trail (E1d), 5,600', May

Sagehen Creek east (N1c), 6,200', May–June

Pole Creek (N4b), 7,400', May–June

Galena Creek (E4e), 7,500', June

5-Lakes Basin (N4e), 6,600', June

NAKED BROOMRAPE
Orobanche uniflora
Broomrape Family (Orobanchaceae)

Occasional in damp meadows, seeps throughout Tahoe and in the Great Basin at low and mid elevation. Single (sometimes 2 or 3) tubular, blue-purple, pink, or yellow flower on short, leafless stem; parasite.

Thomas Creek lower trailhead (E4f), 6,100', May

Sagehen Creek west (N1b), 6,200', June

Pole Creek (N4b), 7,200', July

Sherwood Forest (W1a), 7,800', July

SHOWY PENSTEMON
Penstemon speciosus
Figwort Family (Scrophulariaceae)

Common on dry flats and slopes, rocky ledges throughout Tahoe and in the Great Basin at low, mid, and high elevation to above timberline. Large, deep blue flowers mostly on one side of the stem; leathery leaves with waxy, blue-green surface.

Old Geiger Grade (E1c), 5,100', May

Peavine Mt. east (E1a), 5,500', May–June

Topaz Lake to Markleeville (S2e), 7,000', June

Mt. Rose (E4c), 9,000', June-July

Castle Peak east (N2b), 8,000', July

above Winnemucca Lake (S3b), 9,500', July

NORTHERN BOG VIOLET
Viola nephrophylla
Violet Family (Violaceae)

Occasional in wet, grassy meadows, streambanks in northern and eastern Tahoe and in the Great Basin at low and mid elevation. Heart-shaped or kidney-shaped, toothed, basal leaves; deep purple flower with white throat; lower middle petal narrower than other 2 lower petals.

Peavine Mt. east (E1a), 6,600', May–June

Sagehen Creek west (N1b), 6,200', June

Donner Pass Road (N3a), 7,000, June

PURPLE MILKWEED
Asclepias cordifolia
Milkweed Family (Asclepiadaceae)

Occasional on dry, sandy slopes in eastern and northern Tahoe and in the Great Basin at low and mid elevation. Tall, robust, very leafy plant; umbels of red-purple flowers with strongly reflexed petals above which project purple or white horns resembling corn kernels.

Faye-Luther trail (E1i), 5,500', May–June

5-Lakes Basin (N4e), 7,000', June–July

SNOWPLANT
Sarcodes sanguinea
Heath Family (Ericaceae)

Common in humus of coniferous forests throughout Tahoe at low and mid elevation. Sticky, clammy, red stalks dense with red tubular flowers; plants often in clusters; no green leaves (saprophyte).

Faye-Luther trail (E1i), 5,100', May–June

Vikingsholm (W1g), 6,300', May–June

Prey Meadows (E2a), 6,800', June

Pole Creek (N4b), 6,500', June

Paige Meadows (W1b), 6,500', June–July

Thomas Creek upper trail (E4g), 7,000', June–July

COBWEB THISTLE/SNOW THISTLE
Cirsium occidentale/C. pastoris
Sunflower Family (Asteraceae)

Occasional on dry, sandy flats and slopes (usually with sagebrush) in southeastern Tahoe and in the Great Basin at low elevation. Tall, robust plant; squarish flower heads; plant covered with silvery-gray, cobwebby mat of hairs.

Deadman's Creek (E1f), 5,100', May–June

east of Woodfords (E3a), 5,000', May–June

Topaz Lake (E1j), 5,100', May–June

Old Geiger Grade (E1c), 5,200', June

Peavine Mt. east (E1a), 5,500', June

Faye-Luther trail (E1i), 5,200', June–July

SIERRA LUPINE/GRAY'S LUPINE
Lupinus grayii
Pea Family (Fabaceae)

Occasional on dry or moist flats and slopes, forest openings throughout Tahoe at low and mid elevation. Subshrub; long racemes of pink-purple flowers with yellow patch on banner; strong, sweet fragrance.

Meeks Bay (W1f), 6,300', May–June

Mt. Tallac (W1i), 7,000', July

PURPLE NAMA
Nama aretioides
Waterleaf Family (Hydrophyllaceae)

Rare, on dry, sandy flats and slopes in the Great Basin at low elevation. Red-purple flowers with white and yellow centers; plants usually grow in large patches.

Peavine Mt. south (E1b), 5,000', May

ASPEN ONION
Allium bisceptrum
Lily Family (Liliaceae)

Common in damp meadows, especially around aspen trees, in eastern Tahoe and in the Great Basin at low and mid elevation. 6 pink or rose-purple tepals; 2 or 3 channeled leaves; pair of fringed crests atop ovary.

Deadman's Creek (E1f), 5,100', May

Peavine Mt. east (E1a), 6,000', June

Monitor Pass (S2e), 8,300', June–July

Thomas Creek upper trail (E4g), 7,000', June–July

Galena Creek (E4e), 7,500', June–July

LEMMON'S ONION
Allium lemmonii
Lily Family (Liliaceae)

Occasional in grassy meadows in eastern and southern Tahoe and in the Great Basin at low and mid elevation. Short plant; many tightly clustered, pink or rose flowers with 6 broad, crowded, pointed tepals; 2 papery bracts under flower umbels, 2 leaves.

south of Markleeville (S2e), 5,700', May

west of Monitor Pass (S2e), 7,500', May–June

Tahoe Meadows (E4b), 8,600', July

FLAT-STEMMED ONION
Allium platycaule
Lily Family (Liliaceae)

Rare (locally common) on dry, volcanic flats and slopes in southern and western Tahoe at mid elevation. Short plant; spherical umbel of 6 red-purple flowers with threadlike, widely separated tepals; 2 long, broad, flat leaves.

east of Monitor Pass (S2e), 7,600', May–June

Sherwood Forest (W1a), 8,200', July

APRICOT MALLOW
Sphaeralcea ambigua
Mallow Family (Malvaceae)

Occasional on dry, sandy slopes, usually with sagebrush, in the Great Basin at low elevation. Large, orange or apricot flowers; crinkly, lobed, gray-green leaves; plant covered with short, irritating hairs.

Topaz Lake (E1j), 5,100', May–June

Peavine Mt. east (E1a), 5,500', May–June

Peavine Mt. south (E1b), 5,000', May–June

LASSEN CLARKIA
Clarkia lassenensis
Evening-Primrose Family (Onagraceae)

Rare, on dry flats in the Great Basin at low elevation. A few pink-purple flowers in the leaf axils; 4 unlobed, clawed petals; flowers close at night; narrow, white-hairy leaves.

Thomas Creek lower trailhead (E4f), 6,000', May

ALPINE SHOOTING STAR
Dodecatheon alpinum
Primrose Family (Primulaceae)

Common in wet meadows, along streams throughout Tahoe at low, mid, and high elevation below timberline. Inside-out flowers hang upside down from arching pedicels; 4 pink petals with white and yellow at base; stems and leaves hairless and glandless.

Sagehen Creek east (N1c), 5,900', May–June

Prey Meadows (E2a), 6,700', May–June

Castle Peak east (N2b), 7,500', June–July

Tahoe Meadows (E4b), 8,600', June–July

Winnemucca Lake trail (S3b), 9,000', July–August

WILD ROSE
Rosa woodsii
Rose Family (Rosaceae)

Common along streams, seeps, ditches throughout Tahoe and in the Great Basin at low and mid elevation. Shrub; large, flat, rose-colored flowers with clump of yellow reproductive parts; deliciously fragrant; thorns.

Steamboat Springs (E1e), 4,600', May–June

Deadman's Creek (E1f), 5,000', May–June

Sagehen Creek east (N1c), 6,200', June

Dollar Hill (N5b), 6,400', June–July

Pole Creek headwaters (N4b), 7,500', July

Thomas Creek upper trail (E4g), 7,500', July

APPLEGATE'S PAINTBRUSH
Castilleja applegatei
Figwort Family (Scrophulariaceae)

Common on dry, often rocky slopes, often with sagebrush, throughout Tahoe and in the Great Basin at low, mid, and high elevation below timberline. Wavy-edged, sticky leaves; red, yellow, or orange, lobed bracts.

Topaz Lake (E1j), 5,100', May

Peavine Mt. east (E1a), 6,000', May–June

Ophir Creek trail (E1d), 5,600', May–June

Thomas Creek upper trail (E4g), 8,000', July

Pole Creek (N4b), 6,500', July

Winnemucca Lake trail (S3b), 8,700', July

YELLOW-OR-PURPLE MONKEYFLOWER (RED)
Mimulus densus
Figwort Family (Scrophulariaceae)

Common on dry flats and slopes in southern and eastern Tahoe and in the Great Basin at low and mid elevation. Large flowers on dwarf plant; red-purple petals with 2 yellow bands; glandular calyx and leaves with pungent odor when crushed.

Ophir Creek trail (E1d), 5,600', May–June

Faye-Luther trail (E1i), 4,900', May–June

Steamboat Springs (E1e), 4,600', May–June

Mt. Rose (E4c), 8,900', June–August

Frog Lake (S3b), 8,900', July–August

Snow Valley Peak trail (E2c), 9,000', July–August

RAYLESS DAISY
Erigeron aphanactis
Sunflower Family (Asteraceae)

Occasional on dry flats in eastern and southern Tahoe and in the Great Basin at low and mid elevation. Golden, discoid, buttonlike flower heads atop short, white-hairy stems.

Steamboat Springs (E1e), 4,600', May–June

Topaz Lake (E1j), 5,100', May–June

Ophir Creek trail (E1d), 5,500', May–June

Thomas Creek upper trail (E4g), 8,700', June–July

Frog Lake (S3b), 8,900', June–July

Snow Valley Peak trail (E2c), 8,800', July–August

DOUGLAS WALLFLOWER
Erysimum capitatum
Mustard Family (Brassicaceae)

Common on sandy or rocky flats and slopes throughout Tahoe and in the Great Basin at low, mid, and high elevation to above timberline. Cylindrical racemes of large, 4-petaled flowers; narrow, wavy-edged leaves; long, narrow, flattened seedpods.

Washoe Valley (E1e), 5,000', April–May

Grass Lake (S2a), 7,500', June

Mt. Rose (E4c), 9,000', June–July

Mt. Tallac (W1i), 9,500', July

Freel Peak (S2b), 10,700', July

GOLDEN PHACELIA
Phacelia adenophora
Waterleaf Family (Hydrophyllaceae)

Occasional (locally common) on dry, sandy flats in the Great Basin at low elevation. Ground-hugging plants that often carpet large areas; narrowly bell-shaped flowers; golden yellow petals with faint, purplish veins.

Peavine Mt. south (E1b), 5,000', May

NORTHERN SUN-CUP
Camissonia subacaulis
Evening-Primrose Family (Onagraceae)

Occasional on dry, sandy flats throughout Tahoe and in the Great Basin at low and mid elevation. Large 4-petaled flowers nestled on rosettes of tonguelike, lobed or wavy-edged leaves.

Kyburz Flat (N1a), 5,900', May–June

Martis Valley (N6a), 6,100', May–June

Goose Meadows (N4a), 6,100', May–June

Hope Valley Wildlife Area (S2c), 7,000', May–June

CALIFORNIA POPPY
Eschscholzia californica
Poppy Family (Papaveraceae)

Occasional on dry, sandy flats, disturbed places in eastern and southern Tahoe and in the Great Basin at low elevation. 4 large, golden yellow or orange petals; blue-green, dissected leaves; enlarged rim atop stem directly under flower.

east of **Woodfords** (E3a), 5,000', May–June

Peavine Mt. south (E1b), 5,000', May–June

west of Monitor Pass (S2e), 5,700', May–June

ALTERED ANDESITE BUCKWHEAT
Eriogonum robustum
Buckwheat Family (Polygonaceae)

Rare (locally common) on dry, andesite soil in the Great Basin at low elevation. Large, round leaves matted with felty hairs; large, greenish yellow flower heads subtended by leaflike bracts.

Old Geiger Grade (E1c), 5,000', May

Geiger Grade (E1c), 6,000', May–June

Peavine Mt. south (E1b), 6,200', June

COMMON YELLOW MONKEYFLOWER

Mimulus guttatus
Figwort Family (Scrophulariaceae)

Common in wet meadows, streambanks, seeps throughout Tahoe and in the Great Basin at low and mid elevation. Tall plant; large flowers often with red spots on petals; shiny, broad, serrated leaves in opposite pairs that (on upper stem) fuse around stem.

Deadman's Creek (E1f), 5,000', May–June

Topaz Lake (E1j), 5,100', May–June

Taylor Creek (W1j), 6,300', May–June

Sagehen Creek west (N1b), 6,200', June–July

Winnemucca Lake trail (S3b), 9,000', July–August

MOUNTAIN VIOLET

Viola purpurea
Violet Family (Violaceae)

Common in dry, forest openings throughout Tahoe and in the Great Basin at mid elevation. Round or oval, deeply veined leaves; back of upper 2 petals with large, brown-purple splotches.

Peavine Mt. east (E1a), 6,800', May–June

5-Lakes Basin (N4e), 6,600', May–June

Paige Meadows (W1b), 6,500', July

Grass Lake (S2a), 7,500', July

Thomas Creek upper trail (E4g), 7,500', July

Winnemucca Lake trail (S3b), 8,800', July

BLEPHARIPAPPUS

Blepharipappus scaber
Sunflower Family (Asteraceae)

Occasional on sandy flats in southern Tahoe and in the Great Basin at low elevation. 3-lobed rays; black-purple anthers; feathery, purple styles; often grows in masses.

north of Markleeville (S2d), 5,500', May

Peavine Mt. south (E1b), 5,500', May

Thomas Creek lower trailhead (E4f), 6,100', May–June

GREEN-LEAF MANZANITA
Arctostaphylos patula
Heath Family (Ericaceae)

Common on dry slopes, forest openings throughout Tahoe and in the Great Basin at low and mid elevation. Tall shrub forming dense thickets; smooth, red branches; urn-shaped, white or pink flowers; small "apple" fruits.

Sagehen Creek east (N1c), 5,900', May–June

Vikingsholm (W1g), 6,500', May–June

Ophir Creek trail (E1d), 5,500', May–June

Galena Creek (E4e), 7,500', June

Mt. Tallac (W1i), 7,500', June–July

SOFT LUPINE
Lupinus malacophyllus
Pea Family (Fabaceae)

Occasional on dry, sandy flats in the Great Basin at low elevation. Short plant; soft-hairy leaves; creamy white flowers with yellow or blue tinges.

Peavine Mt. south (E1b), 6,000', May–June

CALIFORNIA HESPEROCHIRON
Hesperochiron californicus
Waterleaf Family (Hydrophyllaceae)

Occasional in damp meadows in southern and northern Tahoe and in the Great Basin at low and mid elevation. Large, white or pink flowers sit directly on rosettes of tonguelike leaves.

Washoe Valley (E1e), 5,000', May

Peavine Mt. east (E1a), 6,800', May–June

Kyburz Flat (N1a), 5,900', May–June

Hope Valley Wildlife Area (S2c), 7,000', May–June

Frog Lake (S3b), 8,800', June–July

SEGO LILY
Calochortus bruneaunis
Lily Family (Liliaceae)

Common on dry flats and slopes, usually with sagebrush, in eastern Tahoe and in the Great Basin at low and mid elevation. 3 petals with maroon-tinged nectar glands overarched with maroon chevrons; maroon anthers; pink stigma.

Topaz Lake (E1j), 5,100', May

Peavine Mt. east (E1a), 6,000', May

Deadman's Creek (E1f), 5,100', May–June

Faye-Luther trail (E1i), 5,000', May–June

Thomas Creek upper trail (E4g), 7,500', July

TUFTED EVENING-PRIMROSE
Oenothera caespitosa
Evening-Primrose Family (Onagraceae)

Occasional on dry, sandy flats in the Great Basin at low elevation. Very large flowers on short stems only slightly above rosette of narrow, wavy-edged leaves; reflexed, pink sepals; 4 petals dry pink or rose-purple after one night of blooming.

Topaz Lake (E1j), 5,100', May

Peavine Mt. south (E1b), 5,000', June

PRICKLY POPPY
Argemone munita
Poppy Family (Papaveraceae)

Common on dry, sandy flats and slopes in the Great Basin at low and mid elevation. 6 large, crinkly petals; scores of bright yellow stamens surrounding a globular, black or dark purple stigma; prickly leaves and stems.

Old Geiger Grade (E1c), 5,000', May–June

Deadman's Creek (E1f), 5,000', May–June

Topaz Lake (E1j), 5,100', May–June

east of Markleeville (S2e), 7,000', May–July

Thomas Creek upper trail (E4g), 7,000', July

Peavine Mt. east (E1a), 7,000', July

STEAMBOAT SPRINGS BUCKWHEAT
Eriogonum ovalifolium ssp. *williamsiae*
Buckwheat Family (Polygonaceae)

Rare (locally common) endemic to Steamboat Springs area (Great Basin) on dry flats at low elevation. Small umbels of cream-colored flowers with red veins on naked stems rising above dense, rounded cushions of blue-green, spatulate, wavy-edged, mostly erect leaves.

Steamboat Springs (E1e), 4,600', May–June

BISTORT
Polygonum bistortoides
Buckwheat Family (Polygonaceae)

Common in wet meadows and bogs, streambanks thoughout Tahoe at low, mid, and high elevation below timberline. Crepe-papery "thumbs" of white or pink-tinged flowers with 5 tiny petal-like sepals, pungent "dirty socks" odor; flower heads on tall, leafless stalks rising above basal, grasslike leaves.

Sagehen Creek east (N1c), 5,900', May–June

Taylor Creek (W1j), 6,400', May–June

Hope Valley Wildlife Area (S2c), 7,000', May–July

Goose Meadows (N4a), 6,100', June–July

Paige Meadows (W1b), 7,000', June–July

Tahoe Meadows (E4b), 8,600', June–July

BITTERROOT
Lewisia rediviva
Purslane Family (Portulacaceae)

Occasional (locally common) on rocky flats in eastern Tahoe and in the Great Basin at low and mid elevation. Large, spectacular flowers with many white or pink petals and sepals; clusters of reproductive parts; pink anthers; usually forms large masses of overlapping flowers.

Peavine Mt. east (E1a), 7,000', May–July

Peavine Mt. south (E1b), 6,000', May–July

Old Geiger Grade (E1c), 6,200', May–July

TOBACCO BRUSH
Ceanothus velutinus
Buckthorn Family (Rhamnaceae)

Common on dry slopes thoughout Tahoe and in the Great Basin at low and mid elevation. Evergreen shrub; large, shiny, oval or round, flammable leaves with spicy fragrance; clusters of tiny, sweetly fragrant flowers with clawed petals.

Faye-Luther trail (E1i), 5,500', May

Woodfords to Carson Pass (S3a), 6,000', May

5-Lakes Basin (N4e), 6,500', June–July

Thomas Creek upper trail (E4g), 7,000', June–July

Mt. Tallac (W1i), 7,000', June–July

Peavine Mt. east (E1a), 7,000', June–July

BOG SAXIFRAGE
Saxifraga oregana
Saxifrage Family (Saxifragaceae)

Occasional in bogs and wet meadows throughout Tahoe at low and mid elevation. Stout, fleshy, glandular-sticky, leafless stalk; tight cluster of tiny flowers with widely separated, spatulate petals; orange anthers.

Sagehen Creek east (N1c), 5,900', May–June

Sagehen Creek west (N1b), 6,200', May–June

Thomas Creek upper trail (E4g), 7,000', June–July

Grass Lake (S2a), 7,500', June–July

Tahoe Meadows (E4b), 8,600', June–July

WHITE-FLOWERED KECKIELLA
Keckiella breviflora
Figwort Family (Scrophulariaceae)

Rare (locally common) on dry, rocky flats and slopes in eastern Tahoe and in the Great Basin at low elevation. Shrub; pair or cluster of flowers on long pedicel out of leaf axils; white petals with rose-purple stripes or tints; upper 2 petal lobes form "awning" over reproductive parts.

Topaz Lake (E1j), 5,100', May–June

JUNE

By mid-June in most years, the low valleys to the east of the Tahoe Basin have heated up to near-summer temperatures, and the peak blooming has left on its "climb" up into mid elevations in the Carson Range. For great gardens and fascinating flowers east of the Tahoe Basin, try Peavine Mountain (all the way to the summit at 8,266') or up to about 8,500' on the Faye Luther trail, the Ophir Creek trail, or along the road to Monitor Pass.

On the west side of the lake, almost any hike or drive up toward the Sierra Crest will boast flowers in at least the first mile or so, up to about 7,500', where you can find the typical, early blooming, chapparal-slope associates, and a few less common early bloomers on gravel flats and in the rocks. If you head down toward the lake on the Vikingsholm trail, you may find the not-so-common purple nightshade, Lobb's nama, and Torrey's lotus and the rare sugarstick and yellow-eyed grass.

Along the east side of the lake, June (or late May) is the time to see the rare (in Tahoe) baneberry on the trail down to Prey Meadows. Although most very high elevation sites—near and above timberline—are not yet in bloom, Mount Rose is a spectacular exception. Because of the strong winds that blow most of the snow off the rocky slopes above timberline, those slopes bloom early.

At the south end of the lake, it's probably still too early for the high elevation hikes, but the mid elevation meadows are lushly blooming, and the fascinating bogs and swamps at Grass Lake and Osgood Swamp are beginning to offer their treasures.

BLUE SAILORS
Cichorium intybus
Sunflower Family (Asteraceae)

Occasional on dry, sandy flats, disturbed places in eastern Tahoe and in the Great Basin at low elevation. Tall, ribbed stems; windmill-like flower heads with blunt, toothed rays; coarse, pinnately lobed leaves. Alien.

Washoe Valley (E1e), 5,000', June–July

Woodfords (E3a), 5,500', June–July

JESSICA'S STICKSEED
Hackelia micrantha
Borage Family (Boraginaceae)

Common on open, grassy slopes, forest openings throughout Tahoe at low, mid, and high elevation below timberline. Small, light blue flowers with slightly raised, white ring around throat; seeds with prickles in distinct lines.

Kyburz Flat (N1a), 5,900', June

5-Lakes Basin (N4e), 6,600', June–July

west-side bowl (E4a), 7,800', June–July

Spooner Lake (E2b), 7,000', July

Winnemucca Lake trail (S3b), 8,900', July

Mt. Tallac (W1i), 9,200', July–August

CRISP THELYPODIUM
Thelypodium crispum
Mustard Family (Brassicaceae)

Rare (locally common), in damp, grassy fields in the Great Basin at low elevation. Tall plant with dense spike of small flowers with purple sepals and 4 white or lavender petals; basal rosette of leaves and clasping, ascending stem leaves.

Washoe Valley (E1e), 5,000', June

BACIGALUPI'S DOWNINGIA
Downingia bacigalupii
Bellflower Family (Campanulaceae)

Rare, in moist, grassy meadows in northern Tahoe at low elevation. Short plant; blue-purple flowers with white and yellow throat; long-projecting, hooked, periscope-like anther tube.

Sagehen Creek east (N1c), 5,900', June

TWO-HORNED DOWNINGIA
Downingia bicornuta
Bellflower Family (Campanulaceae)

Rare (locally common), in pond margins, muddy depressions in the Great Basin at low elevation. Short plant with lower stems often decumbent; blue-purple flowers with 2 white-margined, yellow-green spots; purple "nipples" on lower lip at throat, grows in masses.

Washoe Valley (E1e), 5,000', June–July

PORTERELLA
Porterella carnosula
Bellflower Family (Campanulaceae)

Occasional in edges and drying floors of vernal pools throughout Tahoe and in the Great Basin at low and mid elevation. Blue-purple flowers with white and yellow centers; narrow, fleshy leaves; grows in masses.

Martis Valley (N6a), 6,100', June

Kyburz Flat (N1a), 5,900', June

Peavine Mt. east (E1a), 6,000', June

Pole Creek (N4b), 6,300', June–July

Paige Meadows (W1b), 7,000', June–July

BREWER'S LUPINE
Lupinus breweri
Pea Family (Fabaceae)

Common on dry, rocky flats and slopes, forest openings throughout Tahoe at low, mid, and high elevation below timberline. Short racemes of flowers rising slightly above mostly basal, silky-hairy leaves; conspicuous white patch on banner; grows in masses.

Osgood Swamp (S1a), 6,500', June–July

Tahoe Meadows (E4b), 8,600', June–July

Tamarack Lake (E4d), 8,800', June–July

Mt. Rose (E4c), 9,200', June–August

BROAD-LEAF LUPINE
Lupinus polyphyllus
Pea Family (Fabaceae)

Common in wet meadows, streambanks, seeps throughout Tahoe at mid elevation. Tall, robust plant; long racemes dense with whorls of dark blue flowers; broad, rounded leaflets.

Paige Meadows (W1b), 7,000', June–July

Mt. Rose (E4c), 9,000', June–July

west-side bowl (E4a), 7,800', July

Barker Pass south (W1e), 7,200', July–August

Tahoe Meadows (E4b), 8,600', July–August

Winnemucca Lake trail (S3b), 9,000', July–August

LOBB'S NAMA
Nama lobbii
Waterleaf Family (Hydrophyllaceae)

Rare, on gravelly flats and slopes in southern and western Tahoe at low and mid elevation. Mat-forming subshrub with spherical clusters of blue-purple or red-purple flowers; narrow, felty-hairy leaves.

Vikingsholm trail (W1g), 6,600', June

Meeks Bay (W1f), 6,300', June

PURPLE SAGE
Salvia dorrii
Mint Family (Lamiaceae)

Occasional on dry flats and slopes, usually with sagebrush, in the Great Basin at low elevation. Low-growing, aromatic shrub; densely flowered spikes; blue petals and red-purple bracts.

Peavine Mt. east (E1a), 5,500', June

Peavine Mt. south (E1b), 5,200', June

SHOWY POLEMONIUM
Polemonium pulcherrimum
Phlox Family (Polemoniaceae)

Occasional on rocky slopes and ridges throughout Tahoe at mid and high elevation to above timberline. Low plant forming showy mounds; sticky, fragrant, tightly "ladder-rung" leaves resembling stegosaurus tail; blue-purple flowers with yellow centers.

Mt. Rose (E4c), 10,400', June–August

Frog Lake (S3b), 8,900', July–August

Mt. Tallac (W1i), 9,500', July–August

MONKSHOOD
Aconitum columbianum
Buttercup Family (Ranunculaceae)

Common in wet meadows, bogs, streambanks throughout Tahoe at mid elevation. Tall, robust plant; blue-purple petal-like sepals surround and partially conceal cluster of yellow-green reproductive parts; upper sepals form "hood" that in profile resembles duck head.

Shirley Canyon north (N4c), 6,500', June–July

Paige Meadows (W1b), 6,500', June–July

Osgood Swamp (S1a), 6,500', June–July

Castle Peak east (N2b), 7,700', July

Freel Peak trail (S2b), 8,500', July–August

Marlette Lake trail (E2c), 7,800', July–August

TORREY'S BLUE-EYED MARY
Collinsia torreyi
Figwort Family (Scrophulariaceae)

Common on dry flats throughout Tahoe at low and mid elevation. Few opposite pairs of narrow leaves; whorls of relatively large (for a blue-eyed Mary) flowers; usually grows in masses.

Pole Creek (N4b), 6,300', June

Barker Pass south (W1e), 7,300', June–July

Winnemucca Lake trail (S3b), 8,700', July

AZURE PENSTEMON
Penstemon azureus
Figwort Family (Scrophulariaceae)

Occasional on rocky slopes, forest openings, roadsides throughout Tahoe at mid elevation. Large, intensely blue-purple flowers; yellow buds; short, heart-shaped, clasping, blue-green leaves.

Donner Pass Road (N3a), 6,800', June

Shirley Canyon north (N4c), 6,500', June–July

Grass Lake (S2a), 7,500', July

ALPINE PENSTEMON
Penstemon davidsonii
Figwort Family (Scrophulariaceae)

Occasional (locally common) on rocky slopes, ledges, ridges in southern and northeastern Tahoe at high elevation to above timberline. Dwarfed, creeping plant with large, tubular, blue-purple or red-purple flowers; mats of small, oval leaves.

Mt. Rose (E4c), 10,700', June–August

above Winnemucca Lake (S3b), 9,500', July

Freel Peak (S2b), 10,800', July–August

MEADOW PENSTEMON
Penstemon rydbergii
Figwort Family (Scrophulariaceae)

Common in damp, grassy meadows throughout Tahoe at low and mid elevation. Several whorls of flowers spaced along stem; nonglandular inflorescences; grows in masses.

Goose Meadows (N4a), 6,100', June

Hope Valley Wildlife Area (S2c), 7,000', June

Martis Valley (N6a), 6,100', June

Tahoe Meadows (E4b), 8,600', July–August

PURPLE NIGHTSHADE
Solanum xanti
Nightshade Family (Solanaceae)

Occasional on dry flats, forest openings throughout Tahoe at mid elevation. Blue-purple or lavender petals united into a saucer; yellow, "corncob" of reproductive parts.

Vikingsholm trail (W1g), 6,600', June–July

Meeks Bay (W1f), 6,500', July

WANDERING DAISY
Erigeron peregrinus
Sunflower Family (Asteraceae)

Common in grassy meadows, forest openings throughout Tahoe at mid elevation. Pink or lavender rays; glandular-sticky phyllaries; mostly basal leaves.

west-side bowl (E4a), 7,800', June–July

Mt. Rose (E4c), 9,000', June–July

Paige Meadows (W1b), 6,500', June–July

Tamarack Lake (E4d), 8,700', June–July

Pole Creek (N4b), 7,500', June–July

SAINFOIN
Onobrychis viciifolia
Pea Family (Fabaceae)

Rare, in damp or wet, grassy fields and meadows, disturbed places in southern Tahoe and in the Great Basin at low elevation. Dense racemes of rose-colored flowers with red veins; leaves with 15–20 narrow, more-or-less opposite leaflets and 1 terminal leaflet; woody taproot. Alien.

Deadman's Creek (E1f), 5,000', June

Osgood Swamp (S1a), 6,500', July–August

HORSE-MINT
Agastache urticifolia
Mint Family (Lamiaceae)

Common on moist or dry slopes, grassy meadows, forest openings throughout Tahoe at low and mid elevation. Tall plants; "bottlebrush" spikes of flowers; rose or lavender sepals; pink or white petals; triangular, aromatic leaves; usually grows in masses.

Peavine Mt. east (E1a), 6,000', June

east of Monitor Pass (S2e), 7,500', June–July

Paige Meadows (W1b), 6,500', July

Pole Creek headwaters (N4b), 7,600',
July–August

Thomas Creek upper trail (E4g), 7,500',
July–August

Winnemucca Lake trail (S3b), 8,900',
July–August

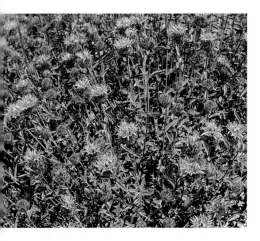

PENNYROYAL
Monardella odoratissima
Mint Family (Lamiaceae)

Common on dry slopes throughout Tahoe and in the Great Basin at low and mid elevation. Roundish heads of flowers with red-purple, blue-purple, pink, or white petals; flowers and especially the opposite leaves have strong mint fragrance.

east of Monitor Pass (S2e), 6,500', June–July

Pole Creek (N4b), 7,100', June–July

Ridge route (N3b), 8,000', June–July

Spooner Lake (E2b), 7,000', July

Barker Peak (W1d), 7,700', July

Winnemucca Lake trail (S3b), 8,900', July–August

THREE-BRACTED ONION
Allium tribracteatum
Lily Family (Liliaceae)

Rare (locally common) on dry, volcanic flats in northern Tahoe at low elevation. 3 (sometimes 2) transparent, whitish bracts under the flower umbel; pink or red-purple (sometimes white) flowers; 2 flat, channeled leaves; grows in masses.

Sagehen Creek east (N1c), 5,900', June

CHECKERMALLOW
Sidalcea glaucescens
Mallow Family (Malvaceae)

Common on dry slopes, often with sagebrush, throughout Tahoe (less common in the south) and in the Great Basin at low and mid elevation. Large, bowl-shaped, pink or rose petals with white veins; palmately lobed leaves mottled with whitish powder.

Peavine Mt. south (E1b), 6,200', June–July

Shirley Canyon north (N4c), 6,200', June–July

east of Monitor Pass (S2e), 7,500', June–July

Paige Meadows (W1b), 6,400', July

Pole Creek (N4b), 7,200', July

Thomas Creek upper trail (E4g), 7,000', July

SCARLET GILIA
Ipomopsis aggregata
Phlox Family (Polemoniaceae)

Common on dry, sandy, or rocky flats and slopes, forest openings throughout Tahoe at low and mid elevation. Tall plant; large, scarlet "trumpets" often with pink or white mottling on petals; needle-lobed leaves.

west-side bowl (E4a), 7,800', June–July

Shirley Canyon north (N4c), 6,300', June–July

Barker Pass south (W1e), 7,500', June–July

Mt. Rose (E4c), 9,000', June–August

Pole Creek (N4b), 7,400', July

Snow Valley Peak trail (E2c), 8,800', July–August

CRIMSON COLUMBINE
Aquilegia formosa
Buttercup Family (Ranunculaceae)

Common in moist meadows, forest openings throughout Tahoe at low, mid, and high elevation below timberline. Tall plant; large, red and yellow, long-spurred flowers hanging upside down; deeply pinnately lobed leaves.

Sagehen Creek west (N1b), 6,200', June

Vikingsholm trail (W1g), 6,400', June

Paige Meadows (W1b), 7,000', June–July

Tamarack Lake (E4d), 8,700', June–July

Winnemucca Lake trail (S3b), 8,900', July–August

OLD MAN'S WHISKERS
Geum triflorum
Rose Family (Rosaceae)

Occasional in moist meadows, forest openings in southern and northern Tahoe at mid and high elevation below timberline. Nodding, "puckered-kiss" flowers; in fruit, delicate cluster of long, sinuous, feathery styles.

Hope Valley Wildlife Area (S2c), 7,000', June

Pole Creek saddle (N4b), 8,400', June–July

Grass Lake (S2a), 7,500', July

Meiss Meadows pond trail (S3c), 9,000', July

Frog Lake (S3b), 8,900', July

PURPLE CINQUEFOIL
Potentilla palustris
Rose Family (Rosaceae)

Rare, in bogs and swampy places in southern Tahoe at mid elevation. Dark wine-red petals; spherical cluster of red-purple reproductive parts; toothed leaflets.

Osgood Swamp (S1a), 6,500', June–July

Grass Lake (S2a), 7,500', July–August

ALPINE PAINTBRUSH
Castilleja nana
Figwort Family (Scrophulariaceae)

Common on sandy or rocky slopes and ridges throughout Tahoe at mid and high elevation to above timberline. Short plant; bracts are muted pink, salmon, gray, or beige.

Winnemucca Lake trail (S3b), 9,000', June–August

Mt. Rose (E4c), 10,700', June–August

Freel Peak (S2b), 10,800', July–August

Pole Creek saddle (N4b), 8,400', July–August

Mt. Tallac (W1i), 9,500', July–August

Red Lake Peak (S3d), 10,000', July–August

COPELAND'S OWL'S-CLOVER
Orthocarpus cuspidatus
Figwort Family (Scrophulariaceae)

Common on dry, sandy flats and slopes, often with sagebrush, in northern and southern Tahoe and in the Great Basin at low and mid elevation. Short plant; rounded bracts with pink-purple tips; flower with projecting pink beak; grows in masses.

Peavine Mt. east (E1a), 6,000', June–July

Peavine Mt. south (E1b), 6,000', June–July

Pole Creek saddle (N4b), 8,400', July

Meiss Meadows pond trail (S3c), 8,800', July

Barker Peak (W1d), 7,600', July

Castle Peak east (N2b), 8,000', July–August

BULL ELEPHANT'S-HEAD
Pedicularis groenlandica
Figwort Family (Scrophulariaceae)

Common in wet meadows, along streams throughout Tahoe at low and mid elevation. Spike of red-purple flowers with "elephant ears" and "trunk" that sweeps down and then curves up; fernlike leaves.

west-side bowl (E4a), 7,800', June

Sagehen Creek west (N1b), 6,200', June

Paige Meadows (W1b), 7,000', June–July

Tahoe Meadows (E4b), 8,600', June–July

Castle Peak east (N2b), 7,500', July

Winnemucca Lake trail (S3b), 9,000', July

MOUNTAIN PRIDE
Penstemon newberryi
Figwort Family (Scrophulariaceae)

Common on rocky slopes, ledges, crevices throughout Tahoe at low and mid elevation. Low-growing shrub; pink or rose flowers; mostly basal, leathery leaves; often grows in masses.

Shirley Canyon south (N4d), 6,500', June

Vikingsholm trail (W1g), 6,400', June

Donner Pass Road (N3a), 6,800', June

Sherwood Forest (W1a), 7,800', June–July

Pole Creek headwaters (N4b), 7,600', July

Frog Lake (S3b), 8,900', July

BIGELOW'S SNEEZEWEED
Helenium bigelovii
Sunflower Family (Asteraceae)

Common in grassy meadows mostly in eastern Tahoe at mid elevation. Tall plant; large flower heads with raised disk and long rays; alternating, clasping leaves well up stem; grows in masses.

Galena Creek (E4e), 8,000', June–July

Thomas Creek upper trail (E4g), 7,200', June–July

Tamarack Lake (E4d), 8,700', June–July

Tahoe Meadows (E4b), 8,700', June–July

ALPINE GOLD
Hulsea algida
Sunflower Family (Asteraceae)

Occasional (locally common) on rocky flats and ridges on only 4 peaks in eastern and southwestern Tahoe at high elevation mostly above timberline. Tall (1' or so) for an alpine plant; large flower heads with many narrow rays; densely glandular-hairy, fragrant, toothed leaves; white-wooly phyllaries.

Mt. Rose (E4c), 10,700', June–August

Mt. Tallac (W1i), 9,700', July–August

Freel Peak (S2b), 10,800', July–August

WOOLY MULE EARS
Wyethia mollis
Sunflower Family (Asteraceae)

Common on dry, volcanic slopes throughout Tahoe (more common in the north) and in the Great Basin at low and mid elevation. Large flower heads; soft-hairy, broad, erect leaves; grows in masses sometimes nearly covering entire slopes.

west-side bowl (E4a), 7,800', June–July

ridge route (N3b), 8,000', June–July

Shirley Canyon north (N4c), 6,300', June–July

Pole Creek (N4b), 7,200', July

Barker Peak (W1d), 7,700', July

Winnemucca Lake trail (S3b), 8,700', July

BIRD'S-FOOT LOTUS
Lotus corniculatus
Pea Family (Fabaceae)

Occasional on dry, sandy flats and slopes, grassy meadows, disturbed places in southern Tahoe at low and mid elevation. Clusters of yellow-orange flowers with red veins on the banner; 5 leaflets; grows in masses. Alien.

north of Markleeville (S2d), 5,500', June

YELLOW-EYED GRASS
Sisyrinchium elmeri
Iris Family (Iridaceae)

Rare, in wet meadows, grassy areas around trees in southern and western Tahoe at low and mid elevation. 6 narrow, yellow-orange tepals; orange anthers; grasslike leaves.

Vikingsholm trail (W1g), 6,400', June

Osgood Swamp (S1a), 6,500', July

BLADDERWORT
Utricularia vulgaris
Bladderwort Family (Lentibulariaceae)

Rare (locally common) in bogs and swampy areas in southern Tahoe at low and mid elevation. Grows in standing water; floating branches trap and digest insects in bladderlike pods; few snapdragon-like flowers along mostly naked stems.

Osgood Swamp (S1a), 6,500', June–August

Grass Lake (S2a), 7,500', July–August

PRETTY FACE
Triteleia ixioides
Lily Family (Liliaceae)

Common on dry, rocky or sandy slopes throughout Tahoe at mid and high elevation below timberline. 6 golden yellow tepals with dark purple midveins; loose umbel of many flowers.

Donner Pass Road (N3a), 7,000', June

Shirley Canyon south (N4d), 6,500', June–July

Galena Creek (E4e), 8,000', June–July

Red Lake Peak (S3d), 9,500', July

Castle Peak west (N2a), 7,500', July

SMOOTHSTEM BLAZING STAR
Mentzelia laevicaulis
Loasa Family (Loasaceae)

Occasional on dry, sandy flats and slopes in eastern Tahoe and in the Great Basin at low elevation. Tall plant; very large starlike flowers with 5 narrow, pointed petals; dense cluster of long, threadlike stamens; saw-toothed leaves; white felty stems and sepals.

Topaz Lake (E1j), 5,100', June–July

Old Geiger Grade (E1c), 5,000', June–July

Peavine Mt. east (E1a), 5,500', June–July

Steamboat Springs (E1e), 4,600', June–July

TANSY-LEAF EVENING-PRIMROSE
Camissonia tanacetifolia
Evening-Primrose Family (Onagraceae)

Common on dry or moist, sandy flats in northern and eastern Tahoe and in the Great Basin at low and mid elevation. Large, 4-petaled flowers nearly flat on ground; deeply lobed leaves; grows in masses.

Washoe Valley (E1e), 5,000', May–June

Kyburz Flat (N1a), 5,900', May–June

OCHRE-FLOWERED BUCKWHEAT
Eriogonum ochrocephalum
Buckwheat Family (Polygonaceae)

Occasional on dry, rocky slopes and ridges in southern and eastern Tahoe and in the Great Basin at mid and high elevation to above timberline. Large, spherical flower heads on different height stems; blue-green, often erect, wavy-edged leaves.

Peavine Mt. east (E1a), 7,000', June

Freel Peak (S2b), 10,800', July

NARROWLEAF MILKWEED
Asclepias fascicularis
Milkweed Family (Asclepiadaceae)

Occasional on dry flats and slopes in eastern Tahoe and in the Great Basin at low elevation. Clusters of small flowers with white or pale pink, reflexed petals and white "horns"; long, narrow leaves.

Topaz Lake (E1j), 5,100', June

Peavine Mt. east (E1a), 5,500', June

east of Monitor Pass (S2e), 5,500', July

LOW CRYPTANTHA
Cryptantha humilis
Borage Family (Boraginaceae)

Occasional on dry, rocky flats and ridges throughout Tahoe and in the Great Basin at mid and high elevation to above timberline. Large, round or cylindrical cluster of many starfishlike flowers; extremely hairy plant.

Peavine Mt. east (E1a), 8,200', June

Mt. Rose (E4c), 10,500', June–July

Red Lake Peak (S3d), 10,000', June–July

Winnemucca Lake trail (S3b), 9,000', July

SNOWBERRY
Symphoricarpos rotundifolius
Honeysuckle Family (Caprifoliaceae)

Occasional in rocky places throughout Tahoe at low, mid, and high elevation below timberline. Shrub with shredding bark; pairs of hanging, tubular flowers; oval leaves; white, pasty berries.

Thomas Creek upper trail (E4g), 7,000', June

Paige Meadows (W1b), 6,800', June–July

Vikingsholm trail (W1g), 6,300', July

Spooner Lake (E2b), 7,100', July

Winnemucca Lake trail (S3b), 8,800', July–August

Mt. Tallac (W1i), 9,700', July–August

CREEK DOGWOOD
Cornus sericea
Dogwood Family (Cornaceae)

Common along streams and lakes throughout Tahoe at low and mid elevation. Shrub with red stems; broad, opposite, dark green, deeply veined leaves; clusters of 4-petaled flowers.

Eagle Lake (W1h), 6,600', June–July

Shirley Canyon north (N4c), 6,300', June–July

Thomas Creek upper trail (E4g), 7,000', July

Donner Pass Road (N3a), 6,700', July

Castle Peak west (N2a), 7,500', July

SUGARSTICK
Allotropa virgata
Heath Family (Ericaceae)

Rare, in coniferous forest in western Tahoe at low and mid elevation. Stout, sticky, white stalk with red stripes; small, white flowers with red anthers; no green leaves (saprophyte).

Vikingsholm trail (W1g), 6,300', June

WHITE CREST LUPINE/WHITE SPURRED LUPINE
Lupinus arbustus
Pea Family (Fabaceae)

Common on dry, open slopes, forest openings in northeastern Tahoe at mid elevation. Dense whorls of white flowers (more common and widespread form of this species has blue flowers); sharp elbow in calyx; narrow, widely separated leaflets.

Thomas Creek upper trail (E4g), 7,000', June–July

Tamarack Lake (E4d), 8,700', June–August

Tahoe Meadows (E4b), 8,600', July–August

Mt. Rose (E4c), 9,000', July–August

DWARF SIERRA ONION
Allium obtusum
Lily Family (Liliaceae)

Occasional on damp, rocky flats mostly in northern and southern Tahoe at mid and high elevation below timberline. Clusters of flowers nearly flat on the ground; 6 white or pink tepals with red or green midveins.

Donner Pass Road (N3a), 6,800', June

Red Lake Peak (S3d), 9,200', July

Frog Lake (S3b), 8,900', July

LEICHTLIN'S MARIPOSA LILY
Calochortus leichtlinii
Lily Family (Liliaceae)

Common on dry flats and slopes throughout Tahoe and in the Great Basin at low, mid, and high elevation below timberline. 3 rounded petals with black or dark maroon chevrons above nectar glands; white anthers.

Faye-Luther trail (E1i), 5,000', June

Donner Pass Road (N3a), 6,800', June

Shirley Canyon north (N4c), 6,500', June

east of Monitor Pass (S2e), 7,600', June–July

Meiss Meadows pond trail (S3c) 8,900', June–July

Castle Peak east (N2b), 7,500', July

Mt. Tallac (W1i), 9,000', July–August

WESTERN TOFIELDIA
Tofieldia occidentalis
Lily Family (Liliaceae)

Rare, in wet meadows and bogs in southwestern Tahoe at mid elevation. Tall, leafless stalk; round head of many tiny, 6-tepaled flowers; grasslike leaves.

Osgood Swamp (S1a), 6,500', June–July

BUCKBEAN
Menyanthes trifoliata
Buckbean Family (Menyanthaceae)

Rare (locally common) in shallow margins of lakes and in bogs in southern Tahoe at mid elevation. Glistening, white, threadlike hairs on the petals; broad leaves in groups of 3.

Osgood Swamp (S1a), 6,500', June–July

Grass Lake (S2a), 7,500', June–July

SIERRA REIN ORCHID
Platanthera leucostachys
Orchid Family (Orchidaceae)

Common in wet meadows, seeps, along streams throughout Tahoe at low and mid elevation. Densely flowered spike; 3 petals and 3 sepals all white; lower petal ends in long, curving spur; long, narrow, grasslike leaves.

Sagehen Creek east (N1c), 5,900', June–July

Tahoe Meadows (E4b), 8,600', June–July

Paige Meadows (W1b), 7,000', June–July

Osgood Swamp (S1a), 6,500', June–July

Pole Creek (N4b), 6,600', June–July

east of Carson Pass (S3a), 7,500', June–July

GRAND COLLOMIA
Collomia grandiflora
Phlox Family (Polemoniaceae)

Common on dry, sandy slopes throughout Tahoe and in the Great Basin at low and mid elevation. Heads of many large, long-tubed, white, orange, salmon, or yellowish flowers; blue anthers; long, narrow, alternating leaves.

Washoe Valley (E1e), 5,000', June

Kyburz Flat (N1a), 5,900', June

Peavine Mt. east (E1a), 6,000', June

west-side bowl (E4a), 7,800', June–July

Barker Peak (W1d), 7,600', July

Castle Peak east (N2b), 8,000', July–August

DRUMMOND'S ANEMONE
Anemone drummondii
Buttercup Family (Ranunculaceae)

Occasional in rocky places, often on edges of melting snow, in northwestern and southwestern Tahoe at mid elevation. Ground-hugging plant; variable number of overlapping, white (blue-tinged), petal-like sepals; finely dissected leaves.

Castle Peak west (N2a), 7,800', June

Castle Peak east (N2b), 8,500', June

Pole Creek headwaters (N4b), 7,400', June–July

PUSSYPAWS
Calyptridium umbellatum
Purslane Family (Portulacaceae)

Common on dry, sandy or gravelly flats and slopes throughout Tahoe at low, mid, and high elevation below timberline. Fuzzy, round heads of crepe-papery, white, pink, or rose flowers on radiating pedicels that hug the ground when it's cold and rise up when it's hot; rosette of leathery, spoon-shaped leaves.

Old Geiger Grade (E1c), 5,000', June–July

Kyburz Flat (N1a), 5,900', June–July

west-side bowl (E4a), 7,800', June–July

Shirley Canyon north (N4c), 6,300', June–July

Galena Creek (E4e), 7,500', July

Mt. Tallac (W1i), 9,700', July–August

ALUMROOT
Heuchera rubescens
Saxifrage Family (Saxifragaceae)

Occasional on rock cliffs and ledges throughout Tahoe at low, mid, and high elevation below timberline. Plumes of tiny flowers on delicate, wispy, leafless stems; flowers white with pink tinges (sepals); scalloped, basal leaves.

Vikingsholm trail (W1g), 6,400', June

Eagle Lake (W1h), 6,600', June

Pole Creek headwaters (N4b), 7,700', July

Carson Pass (S3a), 8,400', July–August

above Winnemucca Lake (S3b), 9,400', July–August

SACRED DATURA/JIMSON WEED
Datura wrightii
Nightshade Family (Solanaceae)

Rare, on dry, sandy flats in the Great Basin at low elevation. Robust plant with enormous (5" long), tubular flowers; large, dark green, triangular leaves. Alien.

Steamboat Springs (E1e), 4,600', June

JULY

With the arrival of July, peak summer blooming reaches the Tahoe Basin. By late July even the highest elevations above timberline will be in at least partial bloom. Only the eastern sites of Carson City, the Thomas Creek lower trail, Jack's Valley Habitat Management Area, and Steamboat Springs will be largely devoid of blooms.

The low-lying meadows to the north of the lake will be past their peak, so for the most spectacular gardens, head up to the mid or high elevation sites, such as the rocks and grassy slopes above Meiss Meadows pond toward the summit of Red Peak Lake. When you're south of Lake Tahoe in July, be sure also to visit the fascinating swamp environments of Osgood Swamp and Grass Lake, where you'll find many flowers in full bloom that occur in few or no other places in Tahoe.

In July the mid elevation meadows and bogs and seeps come into their full glory, as do the highest elevations—in the south, Freel Peak and in the north, Mount Rose (peaking at, respectively, 10,881' and 10,776'). These summits will require more hiking (10 miles or so roundtrip), but what a reward lies in wait for you here in these high lands above timberline—amazing gardens of dwarfed alpine plants bringing dramatic color and form to the harsh, dry, windswept, alpine terrain. Few plants here reach above 6–8", and many are much shorter, barely rising above dense mats or cushions of basal leaves. Most plants here are densely hairy, apparently to create miniature eddies to protect the plant tissue from dessicating wind and intense solar radiation.

BLUEBELLS
Mertensia ciliata
Borage Family (Boraginaceae)

Occasional in wet meadows, seeps in southern, eastern, and northern Tahoe at mid and high elevation below timberline. Tall plant; broad leaves; tubular flowers hang pendant in clusters; flower tube constricted at base.

Spooner Lake (E2b), 7,000', July

Winnemucca Lake trail (S3b), 8,900', July

Meiss Meadows pond trail (S3c), 8,800', July

Red Lake Peak (S3d), 9,200', July

LOBB'S LUPINE
Lupinus lepidus var. *lobbii*
Pea Family (Fabaceae)

Occasional on dry flats, slopes, ridges in southern and northern Tahoe at mid and high elevation below timberline. Short plant; dense, white hair on stems and leaves; large white patch on banner.

Red Lake Peak (S3d), 9,200', July

Pole Creek saddle (N4b), 8,400', July

Winnemucca Lake trail (S3b), 9,000', July

IDAHO BLUE-EYED GRASS
Sisyrinchium idahoense
Iris Family (Iridaceae)

Occasional in wet, grassy meadows throughout Tahoe at low and mid elevation. Star-shaped flower with 6 deep blue-purple tepals; yellow "eye" in throat; grasslike leaves.

Sagehen Creek west (N1b), 6,200', July

Osgood Swamp (S1a), 6,500', July

Thomas Creek upper trail (E4g), 7,500', July

Freel Peak trail (S2b), 8,600', July–August

WHORLED PENSTEMON
Penstemon heterodoxus
Figwort Family (Scrophulariaceae)

Occasional in grassy meadows, rocky slopes, ridges throughout Tahoe at mid and high elevation below timberline. Tall plant in low elevation meadows, dwarfed plant at high elevation; whorled, blue-purple flowers usually with red-purple tubes; glandular inflorescence.

Paige Meadows (W1b), 6,500', July

Pole Creek headwaters (N4b), 7,500', July

Mt. Tallac (W1i), 9,500', July

above Winnemucca Lake (S3b), 9,400', July–August

GLAUCOUS LARKSPUR
Delphinium glaucum
Buttercup Family (Ranunculaceae)

Common in wet meadows, bogs throughout Tahoe at mid elevation. Tall, robust plant crowded with flowers; whitish film on lower part of stem; large, maplelike leaves.

Paige Meadows (W1b), 7,000', July

Barker Pass south (W1e), 7,300', July–August

Tamarack Lake (E4d), 8,700', July–August

Mt. Rose (E4c), 9,100', July–August

Freel Peak trail (S2b), 8,900', July–August

ANDERSON'S THISTLE
Cirsium andersonii
Sunflower Family (Asteraceae)

Occasional on dry flats, forest openings throughout Tahoe at mid elevation. Tall plant; slender flower head of pink or rose disk flowers; shiny, dark green leaves with small spines.

Barker Pass south (W1e), 7,500', July

Thomas Creek upper trail (E4g), 7,200', July

Pole Creek (N4b), 7,500', July–August

Grass Lake (S2a), 7,500', July–August

Marlette Lake trail (E2c), 7,800', July–August

INDIAN BLANKET
Gaillardia aristata
Sunflower Family (Asteraceae)

Rare, in grassy fields in eastern Tahoe at mid elevation. Spectacular multicolored flower head of rose-purple or brown disk flowers, and ray flowers that are red-purple at the base and yellow at the tips; narrow, slightly pinnately lobed leaves.

Galena Creek (E4e), 7,500', July

ROSY SEDUM
Sedum roseum
Stonecrop Family (Crassulaceae)

Occasional on moist, rocky slopes, cliffs throughout Tahoe at mid and high elevation below timberline. Dense, shrublike growth form; succulent leaves; 4-petaled, wine-red flowers.

Sherwood Forest (W1a), 7,500', July–August

Mt. Rose (E4c), 9,500', July–August

Pole Creek headwaters (N4b), 7,700', July–August

Castle Peak east (N2b), 8,000', July–August

above Winnemucca Lake (S3b), 9,300', July–August

RED HEATHER
Phyllodoce breweri
Heath Family (Ericaceae)

Common in wet meadows, rocky ledges, forest openings throughout Tahoe at mid and high elevation below timberline. Evergreen shrub with needle-like leaves; pink or rose, cup-shaped flowers; long-protruding stamens.

Castle Peak east (N2b), 8,000', July

Mt. Tallac (W1i), 8,000', July

above Winnemucca Lake (S3b), 9,200', July–August

BOG WINTERGREEN
Pyrola asarifolia
Heath Family (Ericaceae)

Occasional in moist humus of coniferous forests, seeps throughout Tahoe at mid elevation. Nodding pink flowers; rounded, shiny, basal leaves.

Osgood Swamp (S1a), 6,500', July

Castle Peak east (N2b), 8,500', July

Mt. Tallac (W1i), 7,300', July–August

Thomas Creek upper trail (E4g), 7,400', August

SHASTA CLOVER
Trifolium kingii var. *productum*
Pea Family (Fabaceae)

Occasional in grassy meadows throughout Tahoe at mid and high elevation below timberline. Flowers in upside-down "mop"; red or white petals with red tinges; 3 serrated leaflets.

Red Lake Peak (S3d), 9,200', July

BOG MALLOW
Sidalcea oregana
Mallow Family (Malvaceae)

Common on dry slopes, often with sagebrush, throughout Tahoe (less common in the south) and in the Great Basin at low and mid elevation. Large, bowl-shaped, pink or rose petals with white veins; palmately lobed leaves often mottled with whitish powder.

Hope Valley Wildlife Area (S2c), 7,000', June–July

east of Monitor Pass (S2e), 7,500', June–July

Paige Meadows (W1b), 6,400', July

east of Carson Pass (S3a), 8,000', July

Pole Creek (N4b), 7,200', July

Thomas Creek upper trail (E4g), 7,000', July

ALPINE FIREWEED
Epilobium latifolium
Evening-Primrose Family (Onagraceae)

Rare, on rocky slopes, ledges in southern Tahoe at high elevation below timberline. Flowers similar to fireweed, but low, spreading plant; broader, felty leaves; conspicuous veins on petals.

above Winnemucca Lake (S3b), 9,300', July

ROCK FRINGE
Epilobium obcordatum
Evening-Primrose Family (Onagraceae)

Occasional on rocky slopes and ridges, ledges throughout Tahoe at mid and high elevation below timberline. Large, rose-colored flowers nearly on ground; heart-shaped petals; often grows in large clusters around rocks.

Pole Creek headwaters (N4b), 7,400', July–August

Mt. Rose (E4c), 9,800', July–August

Sherwood Forest (W1a), 8,000', July–August

above Winnemucca Lake (S3b), 9,400', July–August

Mt. Tallac (W1i), 9,700', July–August

PINK GILIA
Ipomopsis tenuituba
Phlox Family (Polemoniaceae)

Occasional on dry, sandy, or rocky flats and slopes, forest openings throughout Tahoe at mid and high elevation below timberline. Same as scarlet gilia (and considered by some botanists to be variety of *I. aggregata*) but with pale pink or white flowers.

Castle Peak east (N2b), 7,800', July

Red Lake Peak (S3d), 9,800', July

Winnemucca Lake trail (S3b), 9,000', July–August

LOBB'S BUCKWHEAT
Eriogonum lobbii
Buckwheat Family (Polygonaceae)

Occasional on dry, rocky slopes throughout Tahoe at mid and high elevation to above timberline. Round, rose, pink, yellow, or beige flower heads radiate on prostrate stems from cluster of spatulate, felty, basal leaves.

Shirley Canyon south (N4d), 6,800', July

ridge route (N3b), 8,000', July–August

Freel Peak (S2b), 10,850', July–August

SIERRA PRIMROSE
Primula suffrutescens
Primrose Family (Primulaceae)

Occasional (locally common) on rocky slopes, ledges, cliffs in northwestern and southern Tahoe at mid and high elevation below timberline. Short plant; pink or rose, pinwheel flowers with yellow throat; leathery, serrated, spoon-shaped leaves; usually grows in masses.

Pole Creek headwaters (N4b), 7,800', July

Castle Peak west (N2a), 8,300', July

Castle Peak east (N2b), 8,000', July

ridge route (N3b), 8,000', July–August

above Winnemucca Lake (S3b), 9,400', July–August

GREAT RED PAINTBRUSH
Castilleja miniata
Figwort Family (Scrophulariaceae)

Common in grassy meadows, along streams, forest openings throughout Tahoe at low and mid elevation. Tall plant; large spikes of bright red bracts; leaves unlobed or with a few, short, very narrow lobes.

Thomas Creek upper trail (E4g), 8,000', July

Sherwood Forest (W1a), 8,100', July

west-side bowl (E4a), 7,800', July–August

Winnemucca Lake (S3b), 9,000', July–August

Pole Creek (N4b), 6,600', July–August

Marlette Lake trail (E2c), 8,000', July–August

LEWIS MONKEYFLOWER
Mimulus lewisii
Figwort Family (Scrophulariaceae)

Common in wet meadows, along streams throughout Tahoe at mid elevation. Tall plant; large flowers with square-tipped, rose or pink petals with dark red-purple veins and yellow throat; often grows in masses.

road to Barker Pass (W1c), 6,800', July

Castle Peak east (N2b), 7,300', July

Galena Creek (E4e), 8,000', July–August

Freel Peak (S2b), 9,000', July–August

Mt. Rose (E4c), 9,000', July–August

Mt. Tallac (W1i), 7,100', July–August

LITTLE ELEPHANT'S-HEAD
Pedicularis attollens
Figwort Family (Scrophulariaceae)

Occasional in wet meadows throughout Tahoe at mid and high elevation below timberline. Pink flowers with "trunk" that curves sideways; narrow, fernlike leaves; glandular, hairy inflorescence.

Sherwood Forest (W1a), 7,500', June–July

Red Lake Peak (S3d), 9,200', July

Tahoe Meadows (E4b), 8,600', July

Mt. Tallac (W1i), 9,200', July–August

Winnemucca Lake (S3b), 9,000', July–August

BRIDGE'S PENSTEMON
Penstemon rostriflorus
Figwort Family (Scrophulariaceae)

Rare, on rocky slopes and banks in southeastern Tahoe at mid elevation. Tall plant; narrowly tubular, bright red flowers with conspicuously reflexed lower lip.

east of Monitor Pass (S2e), 6,700', July

SEEP-SPRING ARNICA
Arnica longifolia
Sunflower Family (Asteraceae)

Common on moist flats and slopes throughout Tahoe at mid and high elevation below timberline. Large, "sunny" flower heads; long, narrow, slightly toothed, rough-hairy leaves in 5–7 opposite pairs.

Sherwood Forest (W1a), 8,000', July

Mt. Rose (E4c), 9,500', July

Mt. Tallac (W1i), 9,700', July–September

SIERRA STONECROP
Sedum obtusatum
Stonecrop Family (Crassulaceae)

Common on rocky slopes, cliffs, ledges throughout Tahoe at mid elevation. Bright yellow, star-shaped flowers; spoon-shaped, succulent stem leaves and basal leaves that turn red with age.

Shirley Canyon north (N4c), 6,400', July

Donner Pass Road (N3a), 6,800', July

Grass Lake (S2a), 7,500', July

Pole Creek saddle (N4b), 8,400', July

Mt. Tallac (W1i), 9,000', July

KELLEY'S TIGER LILY
Lilium kelleyanum
Lily Family (Liliaceae)

Rare, in wet or damp meadows in the Great Basin at low elevation. Tall plant; 6 yellow-orange tepals with maroon spots; tepals curve back so tips almost touch; flowers are pendant on undulating pedicels.

Peavine Mt. east (E1a), 5,800', July

ALPINE TIGER LILY
Lilium parvum
Lily Family (Liliaceae)

Common in wet meadows, seeps, creekbanks throughout Tahoe in low and mid elevation. Tall plant; 6 flaring, orange or red tepals with maroon spots; flowers horizontal or erect on long, straight pedicels.

Osgood Swamp (S1a), 6,500', July

Shirley Canyon north (N4c), 6,300', July

Paige Meadows (W1b), 7,000', July

Castle Peak east (N2b), 7,400', July

Pole Creek (N4b), 6,600', July

WOODY-FRUITED EVENING-PRIMROSE
Oenothera xylocarpa
Evening-Primrose Family (Onagraceae)

Rare, on dry, sandy flats and slopes in northeastern Tahoe at mid and high elevation below timberline. Large flowers nearly on ground; 4 petals turn scarlet and wither after 1 night of blooming; coconut fragrance; broad, red-spotted, basal leaves.

Mt. Rose (E4c), 9,000–9,500', July

Galena Creek (E4e), 8,500', July

YELLOW POND-LILY
Nuphar polysepalum
Water-Lily Family (Nymphaeaceae)

Occasional (locally common) in ponds and in shallows of small lakes throughout Tahoe at low and mid elevation. Large, globelike flowers; large, heart-shaped, fleshy, floating leaves that often choke ponds.

Kybruz Flat (N1a), 5,900', July

Osgood Swamp (S1a), 6,500', July

Grass Lake (S2a), 7,500', July

ALPINE BUTTERCUP
Ranunculus eschscholtzii
Buttercup Family (Ranunculaceae)

Occasional on rocky slopes, ridges throughout Tahoe at high elevation to above timberline. Short stems with 1 terminal, bowl-shaped flower; 5 shiny petals; fanlike leaves; usually grows in clusters.

Mt. Rose (E4c), 10,750', July

above Winnemucca Lake (S3b), 9,200', July–August

FAN-LEAF CINQUEFOIL
Potentilla flabellifolia
Rose Family (Rosaceae)

Occasional in grassy meadows throughout Tahoe at mid and high elevation below timberline. Bright yellow flowers; glossy, dark green, fan-shaped leaves with 3 toothed leaflets.

Winnemucca Lake trail (S3b), 9,000', July–August

Meiss Meadows pond trail (S3c), 8,800', July–August

LEMMON'S KECKIELLA
Keckiella lemmonii
Figwort Family (Scrophulariaceae)

Rare, on dry slopes and banks, forest openings in northwestern Tahoe at mid elevation. Tall plant with several opposite pairs of stem leaves; flowers on long pedicels out of leaf axils; yellow, hooded corollas with red veins.

Pole Creek (N4b), 6,900', July

PRIMROSE MONKEYFLOWER
Mimulus primuloides
Figwort Family (Scrophulariaceae)

Common in wet meadows throughout Tahoe at low, mid, and high elevation below timberline. Ground-hugging plant with small flowers; heart-shaped petals; basal leaves with long hairs (that often hold morning dew).

Tahoe Meadows (E4b), 8,700', July

Osgood Swamp (S1a), 6,500', July

Castle Peak east (N2b), 7,500', July–August

Galena Creek (E4e), 8,000', July–August

Red Lake Peak (S3d), 9,500', July–August

MOTH MULLEIN
Verbascum blattaria
Figwort Family (Scrophulariaceae)

Occasional on dry flats, disturbed places in the Great Basin at low and mid elevation. Tall plant with loose-flowered raceme; purple hairs on filaments; nonwooly leaves. Alien.

Washoe Valley (E1e), 5,000', Jully

WHITE HEATHER/ALPINE HEATHER
Cassiope mertensiana
Heath Family (Ericaceae)

Occasional in wet, rocky areas throughout Tahoe at mid and high elevation below timberline. Delicate, hanging, white bells with red, fingerlike sepals; needle-like, evergreen leaves; grows in masses.

Castle Peak east (N2b), 8,000', July–August

Mt. Tallac (W1i), 8,000', July–August

above Winnemucca Lake (S3b), 9,400', July–August

Mt. Rose (E4c), 10,200', July–August

ALPINE GENTIAN
Gentiana newberryi
Gentian Family (Gentianaceae)

Occasional in grassy meadows throughout Tahoe at mid and high elevation below timberline. Large, tubular flower nearly on ground; green spots on petals; fringes between petals.

Mt. Rose (E4c), 9,000', July–August

Mt. Tallac (W1i), 9,500', July–August

SKULLCAP
Scutellaria californica
Mint Family (Lamiaceae)

Rare, in rocky or grassy places in northern Tahoe at low and mid elevation. Tubular, hooded, "pinch-mouthed" flowers; opposite pairs of usually in-folded leaves.

Shirley Canyon south (N4d), 6,300', July

Shirley Canyon north (N4c), 6,400', July

RIGID HEDGENETTLE
Stachys ajugoides/S. rigida
Mint Family (Lamiaceae)

Occasional in damp, grassy meadows throughout Tahoe at low and mid elevation. Several whorls of flowers spaced along tall stem; petals lined with red-purple spots; opposite pairs of tonguelike, scalloped leaves; hairy stems.

Pole Creek (N4b), 6,600', July

WASHINGTON LILY
Lilium washingtonianum
Lily Family (Liliaceae)

Rare, on dry slopes, ofen with chapparal, forest openings in northern and southwestern Tahoe at mid elevation. Tall plant; large, fragrant flowers; 6 tepals with red spots; so heavily picked that it is now rare in the Tahoe area.

Shirley Canyon south (N4d), 6,400', July

Mt. Tallac (W1i), 7,000', July

Donner Pass Road (N3a), 7,000', July

CORN LILY
Veratrum californicum
Lily Family (Liliaceae)

Common in wet meadows, seeps, creekbanks throughout Tahoe at mid and high elevation below timberline. Tall, robust plants; densely flowered stalks; 6 creamy white tepals with dark green splotch at base; large, distinctly veined leaves; grows in masses. Poisonous.

Paige Meadows (W1b), 7,000', July

Winnemucca Lake trail (S3b), 8,900', July–August

Castle Peak east (N2b), 7,500', July–August

Tahoe Meadows (E4b), 8,600', July–August

Pole Creek (N4b), 6,600', July–August

Mt. Rose (E4c), 9,000', July–August

WHITE LADIES-TRESSES
Spiranthes romanzoffiana
Orchid Family (Orchidaceae)

Occasional in wet, grassy places and bogs throughout Tahoe at low and mid elevation. Densely flowered spike of flowers in twisting "braid"; 3 petals and 3 sepals all white; lower petal broad and drooping.

Sagehen Creek east (N1c), 5,900', July

Grass Lake (S2a), 7,500', July–August

Osgood Swamp (S1a), 6,500', July–August

Tahoe Meadows (E4b), 8,600', July–August

FRINGED GRASS-OF-PARNASSUS
Parnassia fimbriata
Grass-of-Parnassus Family (Parnassiaceae)

Rare, in wet, rocky places in northwest Tahoe at mid elevation. Large flowers with white fringes between petals; yellow-green glands; round, scalloped, basal leaves.

Pole Creek headwaters (N4b), 7,400', July

TOLMIE'S SAXIFRAGE
Saxifraga tolmiei
Saxifrage Family (Saxifragaceae)

Rare, on rocky slopes, ledges, crevices in southern and northern Tahoe at mid and high elevation to above timberline. Mat of succulent, spoon-shaped leaves; clusters of flowers on short stems; noticeably 2-beaked pistil in fruit.

Pole Creek ridge above headwaters (N4b), 8,000', July–August

Freel Peak (S2b), 10,300', July–August

DEER'S TONGUE/MONUMENT PLANT
Swertia radiata
Gentian Family (Gentianaceae)

Occasional (locally common) on dry slopes, forest openings in southern Tahoe at mid and high elevation below timberline. Tall, robust plants with stout stems; amazing number of large flowers with greenish white petals with purple spots and pink nectar glands partly concealed with white or pink hairs; whorls of long, narrow leaves.

Winnemucca Lake trail (S3b), 8,900', July

Meiss Meadows pond trail (S3c), 9,000', July

Red Lake Peak (S3d), 9,500', July

AUGUST

Where July was the explosion, August is the afterglow. In many of our sites, August is a month of transition—from the peak bloom of July to the fading of the flowers, the beginning of the fruits, and the preparation for the cooler, crisper weather of fall. In low elevation sites there will be almost no blooms; while in the higher elevations you will find some late bloomers in peak bloom and most of the earlier bloomers in "afterglow."

Those few species that are only now reaching their peak are such a treat, both for us and for the migrating hummingbirds and the late summer bees. So, although many hikes in August will take you to drying stalks and faded blooms and the beginnings of seeds—pods, siliques, berries, parachutes, burs—some hikes will take you to magnificent blooming, all the more precious for its scarcity. And don't forget about those delicious fruits, many of which will now be ripe—currants, blueberries, strawberries, gooseberries, thimbleberries. (Just be certain, of course, before you eat anything, that you know for certain what it is, for there are some very poisonous fruits as well. It is probably wise that, even for the edible ones, you only eat a few.)

August is a wonderful month of endings and beginnings. Some sites with especially lovely displays of late summer bloomers are: Mount Rose, Castle Peak, Freel Peak, Winnemucca Lake, Red Lake Peak, Pole Creek, Mount Tallac, and Osgood Swamp.

EXPLORER'S GENTIAN
Gentiana calycosa
Gentian Family (Gentianaceae)

Occasional in wet, rocky places, seeps throughout Tahoe at mid and high elevation below timberline. Large, tubular flowers; deep blue petals with green spots and fringes between them; fleshy, opposite, often cupped leaves.

Pole Creek headwaters (N4b), 7,500', August–September

Mt. Tallac (W1i), 7,500', August–September

WESTERN EUPATORIUM
Ageratina occidentalis
Sunflower Family (Asteraceae)

Occasional on rocky slopes, ledges throughout Tahoe at mid and high elevation below timberline. Pink flower heads with disk flowers only; dark green, triangular leaves; grows in masses.

Pole Creek headwaters (N4b), 7,700', August–September

Shirley Canyon south (N4d), 8,000', August–September

Mt. Tallac (W1i), 7,300', August–September

FIREWEED
Epilobium angustifolium
Evening-Primrose Family (Onagraceae)

Common in moist meadows, streambanks, around trees, disturbed places throughout Tahoe at mid elevation. Tall plant; deep rose flowers; 4 separate, clawed petals; leaves turn scarlet or bronze in fall; grows in masses.

Shirley Canyon north (N4c), 6,500', August

Grass Lake (S2a), 7,500', August

Galena Creek (E4e), 8,000', August

Mt. Tallac (W1i), 8,500', August–September

Winnemucca Lake trail (S3b), 8,900', August–September

CALIFORNIA FUCHSIA
Epilobium canum
Evening-Primrose Family (Onagraceae)

Occasional on rocky slopes, ledges, crevices throughout Tahoe at mid elevation. Subshrub thick with flowers; 4 2-lobed petals; long-protruding, scarlet pistil; felty leaves.

Pole Creek headwaters (N4b), 7,800',
August–September

Donner Pass Road (N3a), 7,000',
August–September

east of Carson Pass (S3a), 8,400',
August–October

CANADA GOLDENROD
Solidago canadensis
Sunflower Family (Asteraceae)

Common in wet meadows, along streams mostly in southern and northern Tahoe at low and mid elevation. Tall plant; long, triangular inflorescence of "messy" flower heads with disk and ray flowers intermixed; slightly toothed, pointed stem leaves.

east of Carson Pass (S3a), 7,000', August

Goose Meadows (N4a), 6,100', August

Sagehen Creek east (N1c), 6,200', August

SIERRA CORYDALIS
Corydalis caseana
Poppy Family (Papaveraceae)

Rare (locally common) in wet, rocky places, streambanks in northern Tahoe at mid elevation. Stalks of tubular, spurred flowers heading in all directions off stem; flowers often pink-tinged; grapelike fragrance.

Antone Meadows (N5a), 6,700', August

BROAD-LIPPED TWAYBLADE
Listera convallarioides
Orchid Family (Orchidaceae)

Rare, along streambanks, seeps in southern and eastern Tahoe at low and mid elevation. Short plant; 1 opposite pair of broad, round leaves part way up stem; tiny, greenish white flowers with broadly notched lower petal.

Prey Meadows (E2a), 6,900', August–September

SEPTEMBER

A few of Tahoe's showiest flowers save their peak blooming for this month. Most of these begin blooming in August, but September is the time for some of them to achieve their full splendor. Go back to those cliffs at the headwaters of Pole Creek, to the high meadows of Mt. Tallac, or above Winnemucca Lake and you may find many fall treasures now fully at their peak. These amazing flowers would stand out in any crowd, but they are especially dazzling against the muted background of September. And, of course, we aren't the only ones who appreciate them—the migrating hummingbirds revel in the fuchsia, and last-minute bees "shop" at the gentian and eupatorium "malls."

So, most of the flowers you will see blooming in September have been flowering since at least mid-August, but joining these will be a few species that don't begin their major blooming until this last official month of summer, as if they have been waiting for the spotlight. They may not be the most spectacular of Tahoe's flowers, but they are deeply appreciated as the last bloomers of the year. The scent and feel of change is in the air—an undercurrent of coolness and stillness on even the warmest day. Enjoy the few remaining flowers and the multitudes of seeds as another summer passes.

RUBBER RABBITBRUSH
Chrysothamnus nauseosus
Sunflower Family (Asteraceae)

Common on dry, sagebrush slopes throughout Tahoe and in the Great Basin at low, mid, and high elevation to above timberline. Shrub thick with bright yellow, brushlike, discoid flower heads.

Deadman's Creek (E1f), 5,100', September

Spooner Summit (E2b), 7,000', September

Carson Pass (S3a), 7,500', September

Mt. Rose (E4c), 9,000', September

Freel Peak (S2b), 10,700', September–October

DUCK POTATO/ARUM-LEAF ARROWHEAD
Sagittaria cuneata
Water-Plantain Family (Alismataceae)

Occasional, floating in ponds in eastern, southern, and western Tahoe at low and mid elevation. Floating, arrow-shaped leaves; cluster of flowers slightly above water surface; 3 rounded petals.

Osgood Swamp (S1a), 6,500', September–October

Spooner Lake (E2b), 7,000', September–October

WRIGHT'S BUCKWHEAT
Eriogonum wrightii
Buckwheat Family (Polygonaceae)

Occasional in rocky areas mostly in southern and eastern Tahoe and in the Great Basin at low and mid elevation. 6 petal-like sepals with pink veins and pink undersides; tiny, individual flowers; tangled mat of leaves and stems.

Peavine Mt. south (E1b), 5,000', September

east of Monitor Pass (S2e), 6,000', September–October

Snow Valley Peak trail (E2c), 7,200', September–October

Except for the occasional hanger-on, the flowers have passed for another year. Even the fruits, for the most part, have been and gone, leaving dried leaves, dried stalks, dried pods, and seed casings.

Most of the extensive forests in the Tahoe Basin are coniferous as holding their leaves (needles) throughout the winter makes for a quick start in the spring, and the small surface area and waxy surface of needles minimize evaporation, which is a serious risk in such a dry environment. But in wet or damp meadows, along creeks and ponds and lakes, and in seep areas—in any place where water is near and accessible—Tahoe is blessed with patches and ribbons and, in some places, massive groves of deciduous trees whose leaves burst into flame in late September and October. Acres of aspen joined by willow, alder, ash, cottonwood, and an occasional maple create explosions of gold, yellow, orange, copper, bronze, and red. In many places some of the shrubs and herbs also join the party as their leaves turn various autumn hues. Fireweed, shieldleaf, knotweed, and many others burn scarlet or gold or simmer bronze, yellow, copper, or orange along many of our walks and hikes, so explore and enjoy the colors and aromas of autumn as the plants put on one final show before their long rest.

Although you can find some autumn color on almost any of the strolls, walks, and hikes described in this book, especially spectacular groves (mostly aspen) await you at Marlette and Spooner Lakes, along the road from Woodfords to Carson Pass (especially in Hope Valley), over Monitor Pass, up Pole Creek, Taylor Creek, and Galena Creek, and in Paige Meadows. Take a light jacket, a camera, and your wonder, and walk through an aspen tunnel into the winter to come.

In most years November and December bring heavy blankets of snow, covering all but the taller shrubs and the trees and a few lingering, dead stems of the tallest perennial herbs. Some of these plants projecting above the snow do offer tangible traces of the fall fruits, however—dried, brown seed capsules or brittle, brown pods long split open. You may even find a lone seed or two still attached to the capsule or pod, failed in their attempt to "fly the coop." Other than these scattered remnants of the flower season projecting above the snow, the blooming season lies deeply buried, the many perennials for the most part just roots and residual stems under several feet of snow.

To the east in the low-lying valleys at the base of the Carson Range, most of the winter will be largely snow-free, so brown and gray rather than white will be the primary colors—dried shrubs, including lots of sagebrush and rabbitbrush, with scattered material from herbs between the shrubs decaying back into the ground to feed the next year's sprouting.

OTHER TAHOE BASIN WILDFLOWERS

More than 300 flowers not described in text with, for each, one of the best places and times to see it. An asterisk indicates plants that are rarely or infrequently found in the Tahoe area.

agoseris *(Agoseris, Nothocalais)*
    false *(N. troximoides)* .... Thomas Creek lower trailhead (E4f), 6,000', April–May
    short-beaked *(A. glauca)* .... Peavine Mt. east (E1a), 5,500', May
alfalfa *(Medicago sativa)* .... west of Monitor Pass (S2e), 6,800', July
allophylum, violet *(Allophylum violaceum)* .... Sagehen Creek east (N1c) 6,100', May–June
angelica, Brewer's *(Angelica breweri)* .... Tamarack Lake (E4d), 8,700', July
arnica *(Arnica)*
    heart-leaf *(A. cordifolia)* .... Pole Creek (N4b), 6,300', June
    Sierra *(A. nevadensis)* .... Mt. Rose (E4c), 10,000', July
    soft *(A. mollis)* .... Mt. Rose (E4c), 9,200', July
    twin *(A. sororia)* .... Washoe Valley (E1e), 5,000', June
ash, mountain *(Sorbus californica)* .... Eagle Lake (W1h), 6,600', June–July
aster *(Aster)*
    alpine *(A. alpigenus)* .... Barker Peak (W1d), 8,000', July
    Brewer's golden *(A. breweri)* .... Galena Creek (E4e), 9,000', July
    leafy-headed *(A. foliaceus)* .... Mt. Tallac (W1i), 9,000', July–August
    long-leaf *(A. ascendens)* .... Topaz Lake (E1j), 5,100', July–August
    Rocky Mountain (see long-leaf)
    wavy-leaf *(A. integrifolius)* .... Galena Creek (E4e), 8,000', July
    western *(A. occidentalis)* .... Winnemuccal Lake trail (S3b), 8,900', July–August
astragalus (see locoweed and milkvetch)
avens, large-leaf *(Geum macrophyllum)* .... Sagehen Creek east (N1c), 5,900', June

balsamroot hybrid *(Balsamorhiza hookeri/sagittata)* .... west of Monitor Pass (S2e), 7,000', May–June
baneberry* *(Actaea rubra)* .... Prey Meadows (E2a), 6,800', May–June
bedstraw (see cleavers)
bedstraw, many-flowered *(Galium multiflorum)* .... Deadman's Creek (E1f), 5,200', May
beeplant *(Cleome)*
    broad-podded *(C. platycarpa)* .... Peavine Mt. south (E1b), 5,000', May–June
    yellow *(C. lutea)* .... Peavine Mt. south (E1b), 5,000', June
bindweed, field *(Convolvulus arvensis)* .... Topaz Lake (E1j), 5,100', May–July

bittercress *(Cardamine)*

    Brewer's *(C. breweri)* .... Galena Creek (E4e), 7,500', June–July

    heart-leaf *(C. cordifolia)* .... Pole Creek (N4b), 6,600', June–July

blueberry, western *(Vaccinium uliginosum)* .... Grass Lake (S2a), 7,500', July

blue-eyed Mary, small-flowered *(Collinsia parviflora)* .... Kyburz Flat (N1a), 5,900', June

bouncing bet* *(Saponaria officinalis)* .... east of Woodfords (E3a), 5,100', July

brodiaea, hyacinth *(Triteleia hyacinthina)* .... Paige Meadows (W1b), 7,000', June–July

brooklime, American *(Veronica americana)* .... Sagehen Creek west (N1b), 6,200', June

broomrape *(Orobanche)*

    clustered *(O. fasciculata)* .... Snow Valley Peak trail (E2c), 8,800', July

    corymb *(O. corymbosa)* .... Winnemucca Lake trail (S3b), 9,400', July

buckwheat *(Eriogonum)*

    bear *(E. ursinum)* .... Shirley Canyon north (N4c), 6,500', June–July

    blue mountain *(E. strictum)* .... Peavine Mt. east (E1a), 6,000', June

    Douglas *(E. douglasii)* .... Peavine Mt. east (E1a), 6,500', June

    hoary *(E. incanum)* .... Freel Peak (S2b), 10,850', July–August

    nude *(E. nudum)* .... Sherwood Forest (W1a), 7,100', June–July

    oval-leaf *(E. ovalifolium)* .... Mt. Rose (E4c), 10,400', June–July

    rose *(E. rosense)* .... Mt. Rose (E4c), 10,700', July–August

    spurry *(E. spergulinum)* .... Pole Creek (N4b), 7,500', July

buttercup *(Ranunculus)*

    aquatic *(R. aquatilis)* .... Antone Meadows (N5a), 6,800', June

    western *(R. occidentalis)* .... Taylor Creek (W1j), 6,300', May–June

butterweed *(Senecio)*

    basin *(S. multilobatus)* .... Peavine Mt. south (E1b), 5,500', May

    dwarf mountain *(S. fremontii)* .... Mt. Rose (E4c), 10,000', July–August

camas, death *(Zigadenus venenosus)* .... Meeks Bay (W1f), 6,300', June

catchfly *(Silenc)*

    Douglas *(S. douglasii)* .... Galena Creek (E4e), 7,500', July

    Maguire's *(S. bernardina)* .... Mt. Tallac (W1i), 9,700', July

    Sargent's *(S. sargentii)* .... Freel Peak (S2b), 10,800', July–August

chamaesaracha, dwarf* *(Chamaesaracha nana)* .... Thomas Creek upper trail (E4g), 8,200', July

cherry, bitter *(Prunus emarginata)* .... 5-Lakes Basin (N4e), 6,500', June

chickweed *(Cerastium, Stellaria)*

    meadow *(C. arvense)* .... Thomas Creek upper trail (E4g), 7,500', July

    mountain *(S. longipes)* .... Martis Valley (N6a), 6,100', June

chinquapin, bush *(Castanopsis sempervirens)* .... Osgood Swamp (S1a), 6,500', July

chokecherry, western *(Prunus virginiana)* .... south of Markleeville (S2e), 5,500', May–June

cinquefoil *(Potentilla)*

    Drummond's *(P. drummondii)* .... Frog Lake (S3b), 8,900', June–July five-finger *(P. gracilis)* .... Martis Valley (N6a), 6,100', June

shrubby *(P. fruticosa)* .... Pole Creek saddle (N4b), 8,200', July–August

sticky *(P. glandulosa)* .... Spooner Lake (E2b), 7,000', June–July

cleavers *(Galium aparine)* .... Taylor Creek (W1j), 6,400', May–June

clover *(Melilotus, Trifolium)*

  bowl *(T. cyathiferum)* .... Goose Meadows (N4a), 6,100', May–June

  carpet *(T. monanthum)* .... Castle Peak west (N2a), 7,200', July

  long-stalked *(T. longipes)* .... Martis Valley (N6a), 6,100', May–June

  red *(T. pratense)* .... Woodfords to Markleeville (S2d), 5,500', June

  yellow sweet *(M. officinalis)* .... Woodfords to Markleeville (S2d), 5,500', June–July

  white sweet *(M. alba)* .... east of Monitor Pass (S2e), 5,600', June–July

coralroot, spotted *(Corallorhiza maculata)* .... west-side bowl (E4a), 7,800', June–July

cow parsnip *(Heracleum lanatum)* .... Vikingsholm (W1g), 6,300', June

coyote tobacco *(Nicotiana attenuata)* .... Old Geiger Grade (E1c), 5,000', June–July

creambush *(Holodiscus dumosus)* .... Deadman's Creek (E1f), 5,200', May–June

cress *(Rorippa)*

  Tahoe yellow (see yellow cress, Tahoe)

  western yellow *(R. curvisiliqua)* .... Old Geiger Grade (E1c), 5,000', April–May

cryptantha, common *(Cryptantha affinis)* .... Prey Meadows (E2a), 6,800', June

currant *(Ribes)*

  alpine prickly *(R. montigenum)* .... Pole Creek (N4b), 6,900', June

  wax *(R. cereum)* .... west-side bowl (E4a), 7,800', June–July

daisy *(Erigeron, Leucanthemum)*

  Brewer's *(E. breweri)* .... Meiss Meadow pond trail (S3c), 8,600', July

  Coulter's *(E. coulteri)* .... Paige Meadows (W1b), 7,000', June–July

  cut-leaf *(E. compositus)* .... Mt. Rose (E4c), 10,700', June–July

  dwarf alpine *(E. pygmaeus)* .... Mt. Rose (E4c), 10,400', June–July

  Eaton's *(E. eatonii)* .... west of Monitor Pass (S2e), 7,100', June

  Nevada *(E. nevadincola)* .... Deadman's Creek (E1f), 5,200', April–May

  ox-eye *(L. vulgare)* .... south of Markleeville (S2e), 5,500', June

  rayless cut-leaf *(E. compositus* var. *discoideus)* .... Mt. Rose (E4c), 10,700', July

  thread-leaf *(E. filifolius)* .... Deadman's Creek (E1f), 5,100', April

dandelion *(Agoseris, Taraxacum)*

  common *(T. officinale)* .... Martis Valley (N6a), 6,100', May–June

  orange mountain* *(A. aurantiaca)* .... Thomas Creek upper trail (E4g), 7,500', July

desert plume *(Stanleya pinnata)* .... Peavine Mt. south (E1b), 5,000', May–June

dock, curly *(Rumex crispus)* .... Dollar Hill (N5b), 6,400', June–July

dogbane, spreading *(Apocynum androsaemifolium)* .... Galena Creek (E4e), 7,700', July

draba, Lemmon's *(Draba lemmonii)* .... Mt. Rose (E4c), 10,750', June-August

dusty maiden *(Chaenactis douglasii)* .... Grass Lake (S2a), 7,500', July elderberry *(Sambucus)*

    blue *(S. mexicana)* .... west of Topaz Lake (S2e), 6,500', June–July

    red *(S. racemosa)* .... Pole Creek (N4b), 7,400', July

enchanter's nightshade *(Circaea alpina)* .... Galena Creek (E4e), 7,500', July

evening-primrose—see also suncup *(Camissonia, Oenothera)*

    Hooker's *(O. elata)* .... near Markleeville (S2d), 5,700', June–July

    Nevada *(C. nevadensis)* .... Peavine Mt. south (E1b), 5,000', May

    tall (see Hooker's)

figwort, desert *(Scrophularia desertorum)* .... Topaz Lake (E1j), 5,100', May

filaree *(Erodium cicutarium)* .... Peavine Mt. south (E1b), 5,000', April–May

flower baskets *(Mentzelia congesta)* .... west of Topaz Lake (S2e), 5,200', April–May

forget-me-not, small *(Myosotis laxa)* .... Washoe Valley (E1e), 5,000', June

fritillary, purple *(Fritillaria atropurpurea)* .... Martis Valley (N6a), 6,100', June

gayophytum *(Gayophytum)*

    diffuse *(G. diffusum)* .... Faye-Luther trail (E1i), 6,000', May–June

    lumpy-podded *(G. heterozygum)* .... Snow Valley Peak trail (E2c), 7,500', June–July

geranium, Richardson's *(Geranium richardsonii)* .... Sagehen Creek east (N1c), 6,200', June–July

gilia *(Gilia, Ipomopsis)*

    ball-head *(I. congesta)* .... Pole Creek saddle (N4b), 8,400', July

    Bridge's *(G. leptalea)* .... Donner Pass Road (N3a), 6,800', June–July

    granite (see prickly phlox)

    Great Basin *(G. inconspicua)* .... Carson City (E1g), 4,700', March–April

goldenbush *(Ericameria)*

    single-head *(E. suffruticosa)* .... Snow Valley Peak trail (E2c), 9,000', July–August

    whitestem *(E. discoidea)* .... Mt. Tallac (W1i), 9,700', July–August

goldenrod *(Solidago)*

    alpine *(S. multiradiata)* .... Mt. Rose (E4c), 10,000', July

    northern (see alpine)

gooseberry (Ribes)

    alpine *(R. lasianthum)* .... Sherwood Forest (W1a), 8,000', July

    Sierra *(R. roezlii)* .... Mt. Tallac (W1i), 6,500', July

groundsel, wooly *(Senecio canus)* .... Mt. Rose (E4c), 10,500', June–August

gumweed, curly-cup *(Grindelia squarrosa)* .... west of Topaz Lake (S2e), 5,500', July

hawksbeard *(Crepis)*

    intermediate *(C. intermedia)* .... Peavine Mt. east (E1a), 5,500', June

    long-leaved *(C. acuminata)* .... Topaz Lake (E1j), 5,100', May–June

    western *(C. occidentalis)* .... Old Geiger Grade (E1c), 6,000', May–June

hawkweed *(Hieracium)*

    shaggy *(H. horridum)* .... Freel Peak (S2b), 9,500', July–August

    white *(H. albiflorum)* .... Osgood Swamp (S1a), 6,500', June–July

helianthella, California *(Helianthella californica)* .... Pole Creek (N4b), 7,200', July

hemlock

    Douglas water (see water hemlock, Douglas)

    poison *(Conium maculatum)* .... Washoe Valley (E1e), 5,000', May–June

hesperochiron, dwarf* *(Hesperochiron pumilis)* .... Sherwood Forest (W1a), 7,100', July

hopsage, spiny *(Grayia spinosa)* .... Peavine Mt. south (E1b), 5,000', May–June

horehound *(Marrubium vulgare)* .... north of Markleeville (S2d), 5,500', June

horkelia, dusky *(Horkelia fusca)* .... Meiss Meadows pond trail (S3c), 9,000', July

horsebrush, spineless *(Tetradymia canescens)* .... Topaz Lake (E1j), 5,100', June–July

Indian hemp (see spreading dogbane)

ivesia *(Ivesia)*

    club-moss *(I. lycopodioides)* .... above Winnemucca Lake (S3b), 9,400', July–August

    Gordon's *(I. gordonii)* .... Frog Lake (S3b), 8,900', June–July

    mousetail *(I. santolinoides)* .... Freel Peak (S2b), 8,300', July–August

Jacob's ladder (see polemonium)

jewelflower (see shieldleaf)

kelloggia *(Kelloggia galioides)* .... Pole Creek (N4b), 6,600', July

knotweed *(Polygonum)*

    alpine *(P. phytolaccifolium)* .... Vikingsholm (W1g), 6,300', June

    Douglas *(P. douglasii)* .... Goose Meadows (N4a), 6,100', June

    least *(P. minimum)* .... Tahoe Meadows (E4b), 8,600', July

    Shasta *(P. shastense)* .... Freel Peak (S2b), 10,800', July–August

    subalpine *(P. davisiae)* .... Winnemucca Lake trail (S3b), 8,900', July

labrador tea *(Ledum glandulosum)* .... Osgood Swamp (S1a), 6,500', June

ladies-tresses, yellow *(Spiranthes porrifolia)* .... Osgood Swamp (S1a), 6,500', August

larkspur, Nuttall's *(Delphinium nuttallianum)* .... west of Monitor Pass (S2e), 7,500', May–June

laurel, swamp *(Kalmia polifolia)* .... Mount Rose–north slope (E4c), 10,300', July

lewisia *(Lewisia)*

    Sierra *(L. nevadensis)* .... Paige Meadows (W1b), 6,600', June–July

    three-leaf *(L. triphylla)* .... Paige Meadows (W1b), 6,600', July

linanthus, Nuttall's *(Linanthastrum nuttallii)* .... Freel Peak (S2b), 8,500', July

locoweed *(Astragalus)*

    ballflower *(A. austiniae)* .... Mt. Rose (E4c), 10,400', June–July

    Whitney's *(A. whitneyi)* .... Pole Creek saddle (N4b), 8,400', June

lomatium *(Lomatium)*

    Eaton's giant *(L. dissectum* var. *eatonii)* .... Deadman's Creek (E1f), 5,000', April

    fine-leaf giant *(L. dissectum* var. *multifidum)* .... west of Topaz Lake (S2e), 5,300', May–June

    large-fruited *(L. macrocarpum)* .... Deadman's Creek (E1f), 5,200', April–May

lotus *(Lotus)*

    buck* *(L. crassifolius)* .... Blackwood Canyon (W1c), 6,300', July

    Sierra Nevada *(L. nevadensis)* .... Eagle Lake (W1h), 6,600', June–July

    Spanish *(L. purshianus)* .... Donner Pass Road (N3a), 6,100', June–July

    Torrey's *(L. oblongifolius)* .... Vikingsholm (W1g), 6,600', June–July

lousewort, pinewoods *(Pedicularis semibarbata)* .... west-side bowl (E4a), 7,800', June–July

lovage, Gray's *(Ligusticum grayi)* .... Tamarack Lake (E4d), 8,700', June–July

lupine *(Lupinus)*

    crest (see spurred)

    green-stipuled *(L. fulcratus)* .... Mt. Tallac (W1i), 9,500', July–August

    Nevada *(L. nevadensis)* .... Peavine Mt. east (E1a), 7,500', June

    pine *(L. albicaulis)* .... east of Carson Pass (S3a), 8,000', July

    spurred *(L. arbustus)* .... Shirley Canyon north (N4c), 6,300', June–July

    Tahoe *(L. meionanthus)* .... Snow Valley Peak trail (E2c), 9,000', July–August

    yellow *(L. arbustus* var. *calcaratus)* .... Sagehen Creek west (N1b), 6,200', July

madia (Madia)

    common *(M. elegans)* .... Washoe Valley (E1e), 5,000', June

    least *(M. minima)* .... Sherwood Forest (W1a), 7,100', June

mahagony, curl-leaf mountain *(Cercocarpus ledifolius)* .... Faye-Luther trail (E1i), 5,500', May–June

mallow, common *(Malva neglecta)* .... north of Markleeville (S2d), 5,500', June

manzanita, pine-mat *(Arctostaphylos nevadensis)* .... east of Carson Pass (S3a), 8,000', June–July

mariposa lily (see star tulip)

meadow-rue *(Thalictrum)*

    Fendler's *(T. fendleri)* .... Vikingsholm (W1g), 6,400', June

    few-flowered *(T. sparsiflorum)* .... Antone Meadows (N5a), 6,800', June–July

microseris, nodding *(Microseris nutans)* .... Tamarack Lake (E4d), 8,700', June–July

milkvetch, Humboldt River *(Astragalus iodanthus)* .... Topaz Lake (E1j), 5,100', April

milkweed, showy *(Asclepias speciosa)* .... east of Monitor Pass (S2e), 6,000', July

miner's lettuce *(Montia perfoliata)* .... Faye-Luther trail (E1i), 5,000', April–May

mitrewort, Brewer's *(Mitella breweri)* .... Barker Pass south (W1e), 7,500', June–July

monkeyflower *(Mimulus)*

    Brewer's *(M. breweri)* .... west-side bowl (E4a), 7,800', June–July

    dwarf purple *(M. nanus)* .... Sagehen Creek east (N1c), 6,000', July

    least *(M. leptaleus)* .... west-side bowl (E4a), 7,800', June–July

    mountain *(M. tilingii)* .... above Winnemucca Lake (S3b), 9,200', July–August

    musk *(M. moschatus)* .... Pole Creek headwaters (N4b), 7,800', July reddish *(M. rubellus)* .... Carson City (E1g), 4,700', March–April

Torrey's *(M. torreyi)* .... Paige Meadows (W1b), 6,400', June–July

montia, narrow-leaf *(Montia linearis)* .... Martis Valley (N6a), 6,100', June

mugwort *(Artemisia douglasiana)* .... Pole Creek (N4b), 7,400', July

muilla, Great Basin *(Muilla transmontana)* .... Faye-Luther trail (E1i), 4,900', May

mullein, wooly *(Verbascum thapsus)* .... Topaz Lake (E1j), 5,100', May–June

mustard *(Brassica, Sisymbrium)*

    field *(B. campestris)* .... Washoe Valley (E1e), 5,000', May–June

    tumble *(S. altissimum)* .... Topaz Lake (E1j), 5,100', May–June

nama, Rothrock's *(Nama rothrockii)* .... Mt. Tallac (W1i), 9,700', July

navarretia *(Navarretia)*

    Brewer's *(N. breweri)* .... Goose Meadows (N4a), 6,100', June

    needle *(N. intertexta)* .... Martis Valley (N6a), 6,100', June

nemophila, Sierra *(Nemophila spatulata)* .... west-side bowl (E4a), 7,800', June–July

nettle, stinging *(Urtica dioica)* .... Deadman's Creek (E1f), 5,000', June

onion *(Allium)*

    dark red *(A. atrorubens)* .... Old Geiger Grade (E1c), 6,000', May–June

    Sierra *(A. campanulatum)* .... Meiss Meadows pond trail (S3c), 8,800', June–July

    swamp *(A. validum)* .... Osgood Swamp (S1a), 6,500', June–July

orchid *(Platanthera)*

    Alaska rein *(P. unalascensis)* .... Shirley Canyon south (N4d), 6,500', June

    green bog *(P. sparsiflora)* .... Barker Pass south (W1e), 7,200', July

owl's-clover, hairy *(Orthocarpus tenuis)* .... Kyburz Flat (N1a), 5,900', June

paintbrush *(Castilleja)*

    Lemmon's *(C. lemmonii)* .... Tahoe Meadows (E4b), 8,600', July

    pink *(C. pilosa)* .... Sagehen Creek east (N1c), 5,900', July

parsley, aromatic spring *(Cymopterus terebinthinus)* .... Pole Creek headwaters (N4b), 7,400', July

pea, Nevada *(Lathyrus lanszwertii)* .... Thomas Creek upper trail (E4g), 7,500', June–July

pearly everlasting *(Anaphalis margaritacea)* .... Pole Creek (N4b), 6,900', July

peavine *(Vicia americana)* .... Peavine Mt. east (E1a), 6,000', June

penstemon *(Penstemon)*

    gay *(P. roezlii)* .... Spooner Lake (E2b), 7,000', July

    hot-rock *(P. deustus)* .... south of Markleeville (S2e), 5,500', May–June

    slender *(P. gracilentus)* .... Mt. Rose (E4c), 9,000', June–July

    small-flowered *(P. procerus)* .... Frog Lake (S3b), 8,900', June

peppergrass *(Lepidium)*

    English *(L. campestre)* .... Goose Meadows (N4a), 6,100', June

    round-leaf *(L. perfoliatum)* .... Carson City (E1g), 4,700', March–April phacelia *(Phacelia)*

    ballhead *(P. hydrophylloides)* .... Sherwood Forest (W1a), 7,500', June–July

    branching *(P. ramosissima)* .... Meiss Meadows pond trail (S3c), 8,700', July

    timberline *(P. hastata)* .... Mt. Rose (E4c), 9,000', June–August

    vari-leaf *(P. heterophylla)* .... Deadman's Creek (E1f), 5,000', May–June

phlox *(Leptodactylon, Microsteris, Phlox)*

    compact *(P. condensata)* .... Mt. Rose (E4c), 10,400', June–July

    desert *(P. austromontana)* .... Peavine Mt. east (E1a), 8,200', June

    prickly *(L. pungens)* .... Barker Peak (W1d), 8,150', July

    slender *(M. gracilis)* .... Carson City (E1g), 4,700', March–May

pincushion, alpine Douglas *(Chaenactis douglasii* var. *alpina)* .... Red Lake Peak (S3d), 9,800', July

pinedrops *(Pterospora andromedea)* .... Galena Creek (E4e), 7,500', July

pipsissewa *(Chimaphila menziesii)* .... Eagle Lake (W1h), 6,600', July

plectritis, white *(Plectritis macrocera)* .... Carson City (E1g), 4,700', March–May

podistera, Sierra* *(Podistera nevadensis)* .... Freel Peak (S2b), 10,800', July

polemonium *(Polemonium)*

    great *(P. occidentale)* .... Paige Meadows (W1b), 7,000', June–July

    low *(P. californicum)* .... Grass Lake (S2a), 7,500', June–August

popcorn flower, vernal pool *(Plagiobothrys scouleri)* .... Kyburz Flat (N1a), 5,900', May–June

poverty weed *(Iva axillaris)* .... Peavine Mt. south (E1b), 5,000', June

prickly pear, plains *(Opuntia polyacantha)* .... Peavine Mt. east (E1a), 5,500', June

prince's pine (see pipsissewa)

pteryxia, terebinth (see aromatic spring parsley)

pussytoes *(Antennaria)*

    alpine *(A. media)* .... Mt. Tallac (W1i), 9,700', July

    rosy *(A. rosea)* .... Tamarack Lake (E4d), 8,700', July

    silvery *(A. argentea)* .... Pole Creek (N4b), 6,500', June–July

raillardella *(Raillardella)*

    green-leaf *(R. scaposa)* .... Carson Pass (S3a), 8,500', July

    silver-leaf *(R. argentea)* .... Castle Peak west (N2a), 8,600', July

ranger's buttons *(Sphenosciadium capitellatum)* .... Paige Meadows (W1b), 7,000', July–August

red maids* *(Calandrinia ciliata)* .... Hope Valley Wildlife Area (S2c), 7,000', May–June

rock-cress *(Arabis)*

    blue mountain *(A. puberula)* .... Thomas Creek upper trail (E4g), 8,600', July

    elegant *(A. sparsiflora)* .... west-side bowl (E4a), 7,800', June

    Holboell's *(A. holboellii)* .... Thomas Creek lower trailhead (E4f), 6,000', April–May

    prince's *(A. pulchra)* .... Peavine Mt. south (E1b), 5,500', May–June

sagebrush, big *(Artemisia tridentata)* .... Jack's Valley Habitat Management Area (E1h), 5,200', June

sagewort, boreal *(Artemisia norvegica)* .... Winnemucca Lake trail (S3b), 8,900', July–August

Saint John's wort, Scouler's *(Hypericum formosum)* .... Pole Creek (N4b), 6,600', July

salsify, yellow *(Tragopogon dubius)* .... Sagehen Creek east (N1c), 6,200', June

sandwort *(Arenaria, Minuartia)*

    King's *(A. kingii)* .... Mt. Rose (E4c), 10,400', June–July

    needle-leaf *(A. aculeata)* .... Tamarack Lake (E4d), 8,700', June–July

    Nuttall's *(M. nuttallii)* .... Peavine Mt. south (E1b), 6,000', May–June

saxifrage *(Saxifaga)*

    alpine *(S. aprica)* .... Donner Pass Road (N3a), 6,800', May

    brook *(S. odontoloma)* .... above Winnemucca Lake (S3b), 9,200', July–August

    peak *(S. nidifica)* .... Pole Creek headwaters (N4b), 7,700', June–July

sedum (see also stonecrop)

self-heal *(Prunella vulgaris)* .... Osgood Swamp (S1a), 6,500', June–July

senecio (see also butterweed and groundsel)

senecio, arrow-leaf *(Senecio triangularis)* .... Paige Meadows (W1b), 6,500', June–July

serviceberry *(Amelanchier)*

    glabrous *(A. alnifolia)* .... west of Monitor Pass (S2e), 7,000', May–July

    Utah *(A. utahensis)* .... 5-Lakes Basin (N4e), 6,800', June–July

shepherd's purse *(Capsella bursa-pastoris)* .... Peavine Mt. east (E1a), 5,500', June

shieldleaf *(Streptanthus tortuosus)* .... Castle Peak west (N2a), 7,500', July

shooting star, Jeffrey's *(Dodecatheon jeffreyi)* .... Mt. Tallac (W1i), 8,000', July

sibbaldia* *(Sibbaldia procumbens)* .... Mt. Tallac (W1i), 9,000', July

skeleton plant (see stephanomeria)

snakeroot, Sierra *(Sanicula graveolens)* .... Goose Meadows (N4a), 6,100', May–June

Solomon's seal *(Smilacina)*

    racemose false *(S. racemosa)* .... Sherwood Forest (W1a), 7,300', June–July

    star-flowered false *(S. stellata)* .... Shirley Canyon north (N4c), 6,300', July

sorrel, mountain *(Oxyria digyna)* .... Mt. Tallac (W1i), 9,700', July–August

speedwell, thyme-leaf *(Veronica serpyllifolia)* .... Martis Valley (N6a), 6,100', May–June

spiraea, mountain *(Spiraea densiflora)* .... Castle Peak east (N2b), 7,300', June–July

spring beauty, western *(Claytonia lanceolata)* .... above Winnemucca Lake (S3b), 9,200', July

spurrey, ruby sand *(Spergularia rubra)* .... Sherwood Forest (W1a) . 8,000', July

star tulip, naked* *(Calochortus nudus)* .... Barker Pass south (W1e), 7,000', June

stenotus, short-stemmed *(Stenotus acaulis)* .... Red Lake Peak (S3d), 9,000', July–August

stephanomeria *(Lygodesmia, Stephanomeria)*

    large-flowered *(S. lactucina)* .... Freel Peak (S2b), 9,000', July–August

    narrow-leaf (see wire lettuce)

    thorny *(L. spinosa/S. spinosa)* .... Snow Valley Peak trail (E2c), 8,500', July–August

stickleaf *(Mentzelia)*

    Nevada *(M. dispersa)* .... Sagehen Creek east (N1c), 6,000', June

    whitestem *(M. albicaulis)* .... Peavine Mt. east (E1a), 6,000', May

stickseed, velvety *(Hackelia velutina)* .... Paige Meadows (W1b), 6,400', June–July

violet *(Viola)*

    Macloskey's *(V. macloskeyi)* .... Grass Lake (S2a), 7,500', June

    western dog *(V. adunca)* .... Hope Valley Wildlife Area (S2c), 7,000', July

watercress *(Rorippa nasturtium-aquaticum)* .... Washoe Valley (E1e), 5,000', May

water hemlock, Douglas *(Cicuta douglasii)* .... Peavine Mt. south (E1b), 5,000', July

waterleaf, California *(Hydrophyllum occidentale)* .... Castle Peak east (N2b), 7,300', July

whiskerbrush *(Linanthus ciliatus)* .... Ophir Creek trail (E1d), 5,400', May–June

whitethorn *(Ceanothus cordulatus)* .... 5-Lakes Basin (N4e), 6,500', June–July

white-top *(Cardaria pubescens)* .... Steamboat Springs (E1e), 4,700', May–June

willow, arctic *(Salix arctica)* .... above Winnemucca Lake (S3b), 9,200', July–August

willow-herb *(Epilobium)*

    smooth-stemmed *(E. glaberrimum)* .... Paige Meadows (W1b), 6,500', July

    sticky *(E. glandulosum/E. ciliatum)* .... Tahoe Meadows (E4b), 8,600', July

    Oregon* *(E. exaltatum)* .... Osgood Swamp (S1a), 6,500', July–August

wintercress, American *(Barbarea orthoceras)* .... Martis Valley (N6a), 6,100', May–June

wintergreen *(Pyrola)*

    one-sided *(P. secunda)* .... Castle Peak east (N2b), 7,500', July

    white-veined *(P. picta)* .... Freel Peak (S2b), 9,000', July–August

wire lettuce *(Stephanomeria tenuifolia)* .... Faye-Luther trail (E1i), 5,000', July–August

wishbone bush *(Mirabilis bigelovii)* .... Peavine Mt. south (E1b), 5,000', March–April

wooly marbles, dwarf *(Psilocarphus brevissimus)* .... Martis Valley (N6a), 6,100', June–July

yampah *(Perideridia)*

    Bolander's *(P. bolanderi)* .... Castle Peak east (N2b), 8,000', July

    Parish's *(P. parishii)* .... Kyburz Flat (N1a), 5,900', June–July

yarrow *(Achillea millefolium)* .... Martis Valley (N6a), 6,100', June

yellow cress *(Rorippa)*

    Tahoe* *(R. subumbellata)* .... beach at Meeks Bay (W1f), 6,200', June

    western *(R. curvisiliqua)* .... Old Geiger Grade (E1c), 5,000', April–May

REFERENCES

Blackwell, L. *Wildflowers of the Tahoe Sierra.* Edmonton, Alberta: Lone Pine Publishing,1997.

———. *Wildflowers of the Sierra Nevada and the Central Valley.* Edmonton, Alberta: Lone Pine Publishing, 1999.

———. *Wildflowers of the Eastern Sierra and Adjoining Mojave Desert and Great Basin.* Edmonton, Alberta: Lone Pine Publishing, 2002.

———. *Great Basin Wildflowers.* Guilford, Conn.: The Globe Pequot Press, 2006.

Carville, J. *Hiking Tahoe's Wildflower Trails.* Edmonton, Alberta: Lone Pine Publishing, 2004.

Cronquist, A., A. Holmgren, N. Holmgren, P. Holmgren, and J. Reveal. *Intermountain Flora: Vascular Plants of the Intermountain West* (in seven volumes). New York: New York Botanical Gardens, 1972.

Graf, M. *Plants of the Tahoe Basin.* Berkeley, Calif.: University of California Press, 1999.

Hickman, J., ed. *The Jepson Manual: Higher Plants of California.* University of California Press, Berkeley, California, 1993.

Naclinger, J., F. Peterson and M. Williams. "Peavine Mountain, Nevada, part 2: Vascular Plants." *Mentzelia* 6(2), 1992.

Niehaus, T. and C. Ripper. *A Field Guide to Pacific States Wildflowers.* New York: Houghton Mifflin, 1976.

Shaffer, J. *The Tahoe Sierra: A Natural History Guide to 106 Hikes in the Northern Sierra.* Berkeley, Calif.: Wilderness Press, 1987.

Smith, G. "A Flora of the Tahoe Basin and Neighboring Areas." *The Washington Journal of Biology* 31(1), 1973.

Weeden, N. *Sierra Nevada Flora.* Berkeley, Calif.: Wilderness Press, 1981.

INDEX

ABOUT THE AUTHOR

Laird R. Blackwell and his wife, Melinda, live at almost 8,000' on the eastern escarpment of the Carson Range, deliciously poised midway between the Tahoe Basin to the west and the Great Basin to the east, midway between the alpine summit of Mt. Rose at nearly 11,000' and the low-lying valley of the Reno–Carson City area at about 4,600–5,000'.

In June through August they live surrounded by wild-flowers, but in any month from February to October, a short drive or long walk or hike will take them to blooming—from the first tentative whispers of spring in the lowlands to the last poignant gasp of autumn on the high peaks.

Along with Jungian psychology, ancient mythologies, *Moby Dick,* teaching, the mountains, and (of course) Melinda, wildflowers are closest to Laird's heart and deepest in his soul. What a blessing to live and teach and write and love immersed in the glorious peaks and spectacular gardens of the Tahoe Sierra.